"Raw Data" Is an Oxymoron

Infrastructures Series

edited by Geoffrey Bowker and Paul N. Edwards

Paul N. Edwards, *A Vast Machine: Computer Models, Climate Data, and the Politics of Global Warming*

Lawrence M. Busch, *Standards: Recipes for Reality*

Lisa Gitelman, ed., *"Raw Data" Is an Oxymoron*

Finn Brunton, *Spam: A Shadow History of the Internet*

Nil Disco and Eda Kranakis, eds., *Cosmopolitan Commons: Sharing Resource and Risks across Borders*

"Raw Data" Is an Oxymoron

Edited by Lisa Gitelman

The MIT Press
Cambridge, Massachusetts
London, England

MIT Press books may be purchased at special quantity discounts for business or sales promotional use. For information, please email special_sales@mitpress.mit.edu or write to Special Sales Department, The MIT Press, 55 Hayward Street, Cambridge, MA 02142.

This book was set in Perpetua by Toppan Best-set Premedia Limited, Hong Kong. Printed and bound in the United States of America.

Library of Congress Cataloging-in-Publication Data

"Raw data" is an oxymoron / edited by Lisa Gitelman.
    pages   cm.—(Infrastructures series)
Includes bibliographical references and index.
ISBN 978-0-262-51828-4
1. Information theory. 2. Databases. 3. Data transmission systems. 4. Data warehousing.
I. Gitelman, Lisa, editor of compilation.
Q355.R385 2013
001.4—dc23
2012022465

10 9 8 7 6 5 4 3 2 1

Contents

Acknowledgments

The preparation and publication of *"Raw Data" Is an Oxymoron* has involved sustained collaboration over several years. I'd like to thank Jennie Jackson and the many contributors to the volume, in particular, for their thinking and their patience. Were the zillions of emails we exchanged ever collected into a book, it would be a sizeable companion volume or perhaps a multivolume set. I'd also like to thank series editors Geoffrey Bowker and Paul Edwards for their interest in and support of this project, as well as for their own valuable publications in the emerging field that one might—if one wished—call data studies. Whether it is apparent to them or not, their thinking has been an inspiration at every turn, as have so many ongoing discussions with colleagues in and about the humanities, particularly at the interdisciplinary junction where media studies and science and technology studies share such interesting and important interrogative terrain.

Marguerite Avery at the MIT Press has been an attentive and discerning editor, first to last, as well as a smart, supportive interlocutor, and I'm very grateful to her. Further practical assistance and moral and intellectual support has been rendered by Theresa M. Collins, while my research in the final stages of this project has been supported in part by a fellowship at New York University's Humanities Initiative.

Introduction

Lisa Gitelman and Virginia Jackson

"Raw data" is both an oxymoron and a bad idea.
—Geoffrey C. Bowker, *Memory Practices in the Sciences*

Data are everywhere and piling up in dizzying amounts. Not too long ago storage and transmission media helped people grapple with kilobytes and megabytes, but today's databases and data backbones daily handle not just terabytes but petabytes of information, where *peta-* is a prefix which denotes the unfathomable quantity of a quadrillion, or a thousand trillion. Data are units or morsels of information that in aggregate form the bedrock of modern policy decisions by government and nongovernmental authorities. Data underlie the protocols of public health and medical practice, and data undergird the investment strategies and derivative instruments of finance capital. Data inform what we know about the universe, and they help indicate what is happening to the earth's climate. "Our data isn't just telling us what's going on in the world," IBM advertises; "it's actually telling us where the world is going." The more data the better, by these lights, as long as we can process the accumulating mass. Statisticians are on track to be the next sexy profession in the digital economy, reports the front page of the *New York Times*. "Math majors, rejoice," the newspaper urges in another instance, because businesses are going to need an army of mathematicians as they grapple with increasing mountains of data.[1]

What about the rest of us? What are we to data and data to us? As consumers we tend to celebrate our ability to handle data in association with sophisticated technology. My iPad has 64 gig! My phone is 4G! We don't always know what this means and typically don't know how these devices actually function, but they are "friendly" to users in part according to the ways they empower us to store, manipulate, and transmit data.

Yet if data are somehow subject to us, we are also subject to data, because Google collects so much information on users' interests and behaviors, for instance, and the U.S. National Security Agency mines fiber-optic transmissions for clues about terrorists. Not too long ago it was easier to understand the ways that data was collected about us, first through the institutions and practices of governmentality—the census, the department of motor vehicles, voter registration—and then through the institutions and practices of consumer culture, such as the surveys which told us who we were, the polls which predicted who we'd elect, and the ratings which measured how our attention was being directed. But today things seem different—in degree if not always in kind—now that every click, every move has the potential to count for something, for someone somewhere somehow. Is data about you *yours*, or should it be, now that data collection has become an always-everywhere proposition? Try to spend a day "off the grid" and you'd better leave your credit and debit cards, transit pass, school or work ID, passport, and cell phone at home—basically, anything with a barcode, magnetic strip, RFID, or GPS receiver.[2]

In short, if World War II helped to usher in the era of so-called Big Science, the new millennium has arrived as the era of Big Data.[3] For this reason, we think a book like *"Raw Data" Is an Oxymoron* is particularly timely. Its title may sound like an argument or a thesis, but we want it to work instead as a friendly reminder and a prompt. Despite the ubiquity of the phrase *raw data*—over seventeen million hits on Google as of this writing—we think a few moments of reflection will be enough to see its self-contradiction, to see, as Bowker suggests, that data are always already "cooked" and never entirely "raw." It is unlikely that anyone could disagree, but the truism no more keeps us from valuing data than a similar acknowledgment keeps up from buying jumbo shrimp. The analogy may sound silly, but not as silly as it first appears: just as the economy of shrimp and shrimping has shifted radically in the decades since the birth of industrial aquaculture in the 1970s, so the economy of data has an accelerated recent history. The essays in this volume do not present one argument about that economy, but they do begin to supply a little heretofore-unwritten history for the seismic shift in the contemporary conception and use—the sheer existence—of so much data.

However self-contradicting it may be, the phrase *raw data*—like *jumbo shrimp*—has understandable appeal. At first glance data are apparently before the fact: they are the starting point for what we know, who we are, and how we communicate. This shared sense of starting with data often leads to an unnoticed assumption that data are transparent, that information is self-evident, the fundamental stuff of truth itself. If we're not careful, in other words, our zeal for more and more data can become a faith in their neutrality and autonomy, their objectivity. Think of the ways people talk and write about data. Data are familiarly "collected," "entered," "compiled," "stored," "processed," "mined," and "interpreted." Less obvious are the ways in which the final term in this sequence—interpretation—haunts its predecessors. At a certain level the collection and management of data may be said to presuppose interpretation. "Data [do] not just exist," Lev Manovich explains, they have to be "generated."[4] Data need to be imagined *as* data to exist and function as such, and the imagination of data entails an interpretive base.

Here another analogy may be helpful. Like *events* imagined and enunciated against the continuity of time, *data* are imagined and enunciated against the seamlessness of phenomena. We call them up out of an otherwise undifferentiated blur. If events garner a kind of immanence by dint of their collected enunciation, as Hayden White has suggested, so data garner immanence in the circumstances of their imagination.[5] Events produce and are produced by a sense of history, while data produce and are produced by the operations of knowledge production more broadly. Every discipline and disciplinary institution has its own norms and standards for the imagination of data, just as every field has its accepted methodologies and its evolved structures of practice. Together the essays that comprise *"Raw Data" Is an Oxymoron* pursue the imagination of data. They ask how different disciplines have imagined their objects and how different data sets harbor the interpretive structures of their own imagining. What are the histories of data within and across disciplines? How are data variously "cooked" within the varied circumstances of their collection, storage, and transmission? What sorts of conflicts have occurred about the kinds of phenomena that can effectively—can ethically—be "reduced" to data?

Treating data as a matter of disciplines—rather than of computers, for instance—may seem curious at first. The subject of data is bound to alienate students and scholars in disciplines within the humanities particularly. Few literary critics want to think of the poems or novels they read as "data," and for good reason. The skepticism within literary studies about Franco Moretti's "distant reading" approach, which in part reduces literary objects to graphs, maps, and other data visualizations, testifies to the resistance the notion of literature as data might provoke. Similarly, many historians would not like to reduce their subjects to abstract objects useful in the production of knowledge about the past. Their reluctance was evidenced by the hostile reception accorded to cliometrics in the 1960s and it persists today. In some sense, data are precisely *not* the domain of humanistic inquiry. Yet we propose that students and scholars in the humanities do worry about data, broadly speaking, to the extent that they worry about how their objects of study have been assumed as well as discerned. Don't all questions presuppose or delimit their answers to some degree? Recent work in historical epistemology has challenged the status of the research object, or as Michel Foucault would have it, has raised questions about the boundaries of the archive, about the form, appearance, and regularity of the statements and practices available to us in knowing what we know.[6] When we put our own critical perspectives into historical perspective, we quickly find that there is no stance detached from history, which is to say that there is no persistently objective view.

The conditions of evolving, possessing, and assessing knowledge turn out to be remarkably available to cultural and historical change. The field of science studies has pursued this observation in the greatest detail, and *"Raw Data" Is an Oxymoron* is inspired by science studies while directed beyond it to a broader audience. Evolved over the same decades as other "studies"—like area studies, ethnic studies, cultural and media studies—science studies takes as its object the work of scientists and engineers.[7] The field has helped to confound simplistic dichotomies like theory/practice and science/society in a rich, diverse body of work that, among other things, has explored the situated, material conditions of knowledge production. Looking at the ways scientific knowledge is produced—rather than innocently "discovered," for instance—resembles our project of looking into data or, better, looking *under* data to consider their root assumptions.[8] Inquiries such as these may be seen as contributions toward a critique of objectivity. The point of such a critique—we must quickly emphasize—is not that objectivity is *bad* or that objectivity is mythical. Any such claim must depend, as Lorraine Daston and Peter Galison note, on first achieving a careful understanding of "what objectivity *is*."[9] The point is not how to judge whether objectivity is possible—thumbs up or thumbs down—but how to describe objectivity in the first place. Objectivity is situated and historically specific; it comes from somewhere and is the result of ongoing changes to the conditions of inquiry, conditions that are at once material, social, and ethical.

The very idea of objectivity as the abnegation, neutrality, or irrelevance of the observing self turns out to be of relatively recent vintage. Joanna Picciotto has recently suggested that "the question raised by objectivity is how innocence, traditionally understood to be a state of ignorance, ever came to be associated with epistemological privilege."[10] As a moment in which we can see the emergence of a modern privileging of objectivity, Picciotto nominates "the seventeenth century's conversion of the original subject of innocence, Adam, into a specifically intellectual exemplar. Used to justify experimental science, an emergent public sphere, and the concept of intellectual labor itself," Adam became emblematic of "a new ideal of estranged and productive observation."[11] This means that Milton's *Paradise Lost* and *Paradise Regain'd* may be as important to the development of experimental science as the invention of the microscope.

The innocent observer has had a long, diverse career. Looking at scientific atlases, not Milton poems, Daston and Galison discern the arrival of a version of objectivity that is mechanical: characterized by the observer's restraint and distinguishable from other versions in which the skill and discernment of the observing self counts for something, such as cases in which knowledgeable observers idealize multiple, idiosyncratic specimens into a single type, or in which practiced diagnosticians exert trained judgment in order to make sense of blurry scans. Mechanical objectivity emerged as a dominant ideal in the sciences only in the middle of the nineteenth century, and it is perhaps simplest to describe it contextually with reference to the development of photography during those same years. When Louis Daguerre, Henry Fox Talbot, and others developed and then popularized the first photographic processes, observers were struck by the apparent displacement of human agency in the production of lifelike images. Fox Talbot's lavish account of his calotype process captures this displacement in its title, *The Pencil of Nature*. No artist necessary. Light itself is enough. Photography is objective.

David Ribes and Steven Jackson (chapter 8) direct attention toward some of the difficulties that mechanical objectivity presents today in scientific practice, when biologists rely upon data collected by remote sensors. But mechanical objectivity was something of a conundrum even in Fox Talbot's day. From the very first, the mechanical objectivity of photography was framed by a counter discourse in which photographers were praised for their ability to capture "inner" or "higher" truths on film. The pencil of nature is not enough. Artists are necessary. Photography is subjective. This isn't a question of *either/or* as much as a matter of *and yes*: mechanical objectivity is an "epistemic virtue" among other competing virtues.[12] The presumptive objectivity of the photographic image, like the presumptive rawness of data, seems necessary somehow—resilient in common parlance, utile in commonsense—but it is not sufficient to the epistemic conditions that attend the uses and potential uses of photography. At the very least the photographic image is always framed, selected out of the profilmic experience in which the photographer stands, points, shoots. Data too need to be understood as framed and framing, understood, that is, according to the uses to which they are and can be put. Indeed, the seemingly indispensable misperception that data are ever raw seems to be one way in which data are forever contextualized—that is, framed—according to a mythology of their own supposed decontextualization.

Thus the history of objectivity turns out to be inescapably the history of subjectivity, of the self,[13] and something of the same thing must hold for the concept of data. Data require our participation. Data need us. Yet for all of the suggestive parallels, the history of objectivity is not the history of data. Where did the modern concept of data come from? The first two chapters in this volume tackle this question in different ways. In "Data before the Fact" (chapter 1), Daniel Rosenberg plumbs the derivation and use of *datum* (the singular form) and *data*, offering an intellectual history of the concept that stretches back to the Enlightenment, before the virtue of mechanical objectivity had fully taken shape. Rosenberg is aided in his study—if also provoked—by a new set of tools that offer ways to find and visualize patterns within the digitized corpus of Western printed thought. He gives us the data on data, as it were. Travis D. Williams heads even further back in time, to the Renaissance, in order to consider the history behind one of the strongest epistemic conditions shaping the contemporary data imaginary: the self-evidence of numbers and arithmetic fact as such. Previous scholars have rendered the history of math as or relating to a pre-history of capitalism, and Williams's "Procrustean Marxism and Subjective Rigor" (chapter 2) seeks an additional path, giving an account of English math books with their hilariously prosaic story problems. Like Rosenberg's self-conscious use of present tools in rendering the past, Williams is at pains to take early modern math on its own terms while also considering just what such an endeavor means, since the terms of math are supposed to be universal in time and space. Two plus two equals four, always and everywhere, and "Numbers never lie."

No two chapters could exhaust the multiple origins of data as a concept; Rosenberg and Williams only open the question in different ways. The association of data with diagrams and graphs, in the first instance, and with numbers and mathematical functions, in the second, leads us to the general precept that *data are abstract*. While this quality can make it hard to think or write about data in general—that is, in the abstract—it follows from their abstraction that data ironically require material expression. The retention and manipulation of abstractions require stuff, material things. Just as Cambridge University could become a training ground for mathematical physics only after the introduction of written exams at the end of the eighteenth century (paper and pencil are the things of things where modern abstractions are concerned), so the contemporary era of Big Data has been enabled by the widespread availability of electronic storage media, specifically mainframe computers, servers and server farms, and storage area networks.[14] Both the scale and ontology of electronic storage pose an interesting challenge across the humanities, where lately there has been a renewed interest in *things*.[15] Indeed, as Wendy Hui Kyong Chun has observed, this current scholarly interest in things or "thing theory" needs to be seen against the context of digital media within which things "always seem to be disappearing" in such crucial ways.[16] What sort of things are electronic data, after all?

As we suggested earlier, one productive way to think about data is to ask how different disciplines conceive their objects, or, better, how disciplines and their objects are mutually conceived. The second pair of chapters in this volume takes that tack. In "From Measuring Desire to Quantifying Expectations" (chapter 3), Kevin R. Brine and Mary Poovey address the discipline of economics, and in "Where Is That Moon, Anyway?" (chapter 4), Matthew Stanley considers astronomy. Brine and Poovey follow the work of Irving Fisher, the twentieth-century economist who created the scaffolding for today's financial modeling by linking capital to the concept of present value, which calculates value by taking into account expectations about future yields or benefits. Although the data he used needed to be "scrubbed" to be usable, models like those that Fisher created continue to be influential because they claim a basis that is situated as the objective source of information it can never actually be. As Rosenberg's history helps us understand, this fundamental contradiction may actually be intrinsic to the concept of data, since "the semantic function of data is specifically *rhetorical*." Data by definition are "that which is given prior to argument," given in order to provide a rhetorical basis. (Facts are facts—that is, they are true by dint of being factual—but data can be good or bad, better or worse, incomplete and insufficient.) Yet precisely because data stand as a given, they can be taken to construct a model sufficient unto itself: given certain data, certain conclusions may be proven or argued to follow. Given other data, one would come to different arguments and conclusions.

Disciplines operate according to shared norms, and data scrubbing is an accepted and unexceptional necessity in economics and finance. Disciplines also operate by dint of "data friction"—Paul Edwards's term—friction consisting of worries, questions, and contests that assert or affirm what should count as data, or which data are good and which less reliable, or how big data sets need to be.[17] Stanley's chapter offers a fascinating example of data friction in the field of astronomy. In efforts to derive a particular lunar constant—called the secular acceleration—astronomers have repeatedly engaged in research that on its face seems a lot less like astronomy than it does textual analysis, history, and psychology: poring over the works of classical authors to evaluate their accounts of solar eclipse. The apparent intrusion of psychology into astronomy, or history into climate science, or bibliography into botany—to mention additional examples recently documented—serves as a reminder of just how diverse and dynamic disciplines are.[18] Disciplines aren't just separate subjects you pick out of a course catalogue. They involve infrastructures comprised of "people, artifacts, and institutions that generate, share, and maintain specific knowledge" in complex and interconnected ways.[19] The bodies of knowledge made and maintained by the professions can be more or less specific than those of academic disciplines, but they involve related infrastructures and a similarly evolved and evolving "trust in numbers."[20]

Data aren't only or always numerical, of course, but they do always exist in number in the sense that data are particulate or "corpuscular, like sand or succotash." Something like information, that is, data exist in little bits.[21] This leads us to a second general precept, that *data are aggregative*. They pile up. They are collected in assortments of individual, homologous data *entries* and are accumulated into larger or smaller data *sets*. This aggregative quality of data helps to lend them their potential power, their rhetorical weight. (More is better, isn't it?) Indeed, data are so aggregative that English usage increasingly makes many into one. The word *data* has become what is called a mass noun, so it can take a singular verb. Sentences that include the phrase "data is . . ." are now roughly four times as common (on the web, at least, and according to Google) as those including "data are . . ." despite countless grammarians out there who will insist that *data* is a plural. So far in this introduction we have been assiduous in using the word *data* with plural verbs, and some readers may already have sensed the strain. Data's odd suspension between the singular and the plural reminds us of what aggregation means. If a central philosophical paradox of the Enlightenment was the relation between the particular and the universal, then the imagination of data marks a way of thinking in which those principles of logic are either deferred or held at bay. The singular *datum* is not the particular in relation to any universal (the elected individual in representative democracy, for example) and the plural *data* is not universal, not generalizable from the singular; it is an aggregation. The power within aggregation is relational, based on potential connections: network, not hierarchy.

To be sure, data also depend upon hierarchy. Part of what distinguishes data from the more general category, information, is their discreteness. Each datum is individual, separate and separable, while still alike in kind to others in its set. It follows that the imagination of data is in some measure always an act of classification, of lumping and splitting, nesting and ranking, though the underlying principles at work can be hard to recover. Once in place, classification schemes are notoriously difficult to discern and analyze, since "Good, usable systems disappear almost by definition. The easier they are to use, the harder they are to see."[22] This is the provocation animating an important book by Bowker and Susan Leigh Star entitled *Sorting Things Out*. Working with a group of examples—such as classifying causes of death; classifying the labor of healthcare workers; and classifying race in apartheid-era South Africa—Bowker and Star illuminate the ways that classifications function, for good and ill, to underpin the social order. When phenomena are variously reduced to data, they are divided and classified, processes that work to obscure—or *as if* to obscure—ambiguity, conflict, and contradiction.

Today the ubiquitous structures of data aggregation are computational forms called relational databases. Described and developed since 1970, relational databases organize data into separate tables ("relational variables") in such a way that new data and new kinds of data can be added or subtracted without making the earlier arrangement obsolete. Data are effectively made independent of their organization, and users who perform logical operations on the data are thus "protected" from having to know how the data have been organized.[23] The technical and mathematical details are not important here, but imagine sorting a giant stack of paperwork into separate bins. Establishing which and how many bins are appropriate would be your first important task, but it is likely that as you proceed to sort your papers, you will begin to have a nagging sense that different bins are needed, or that some bins should be combined, or that some papers impossibly belong in multiple bins. You may even wind up with an extra bin or two marked "miscellaneous" or "special problems." It is just this sort of tangle that database architecture seeks to obviate while making relational variables (bins) and their data (papers) available to a multiplicity of desirable logical operations, like queries.

The third pair of chapters in this volume, "facts and FACTS" by Ellen Gruber Garvey (chapter 5) and "Paper as Passion" by Markus Krajewski (chapter 6), takes our paperwork metaphor at face value. Each imagines a different prehistory of the database by considering a specific trove of paper. Garvey describes a giant mass of clippings taken from Southern newspapers to document the horrors of slavery in the antebellum United States, while Krajewski describes the enormous file amassed in the twentieth century by the German sociologist and prolific theorist Niklas Luhmann. Two examples could hardly exhaust the possible prehistories of databases—papery and not—which reach at least as far back as early modern note-taking practices and the accompanying sense of what can anachronistically be called "information overload" that together led to giant compendia with elaborate finding aids.[24] Yet Garvey's example comes from that important moment when the concept of information—close relative of data—finally emerged in something like its present form, as the alienable, abstract contents of an *inform*ative press,[25] while Krajewski's example comes from the equally important moment of systems theory and cybernetics in the second half of the twentieth century.

Garvey's trick, or rather, the trick of the Grimké sisters she writes about, is to fix on an instance where information collected in one locale can take on wholly different meanings in another, as advertisements for runaway slaves become data in the argument against slavery. This is fully remaking the power of the press in the user-dimension, where users may differ in locale if also in their gender, race, and politics. Krajewski by contrast addresses a single user, Niklas Luhmann, who is famous in some quarters for working from his own huge and all-encompassing card index. Author of more than forty books—not a few of them considered "difficult"—Luhmann developed his systems theory, Krajewski suggests, because of, out of, and in collaboration with his card index, a sort of paper machine—a system—for remembering and for generating thought. Papery databases are only metaphorically databases, of course, yet the example of Luhmann's card index helps to clarify the extraordinary generative power that data aggregation can possess while also raising the question of the human or—one must wonder—the posthuman, the human-plus-machine/machine-plus-human hybrids that living with computers make increasingly integral to our understanding.

The final pair of chapters, "Dataveillance and Countervailance" by Rita Raley (chapter 7) and "Data Bite Man" by David Ribes and Steven J. Jackson (chapter 8), pursues the question of data in the present day. Readers will be challenged to think in some detail about the kinds of data being collected about them today, and they will be challenged to consider the difficulties that scientists and policy makers confront when they try to make data useful today and also reusable potentially by others in the future. What are the logics and the ethics of "dataveillance," now that we appear to be moving so rapidly from an era of expanding data resources into an era in which we have become the resource for data collection that vampirically feeds off of our identities, our "likes," and our everyday habits? If while using the Internet we click on a book or a pair of shoes at Amazon.com, or in a box to sign a petition to stop a Congressional bill, or on a link to a porn website, or on a Google Books page or on an online map to find directions, are we making a choice or are we giving Amazon and the federal government and the pornographers (and the security agencies trolling them) and their advertisers ways to guide our choices, calculate our votes, or put us in jail? Both, Raley answers, and suggests that activist projects that exploit dataveillance—that do not opt out but instead "insist on a near-total inhabitation of the forcible frame"—might stand the best chance of at least offering an immanent critique of the predicament that we have created and now must find a way to inhabit.

Ribes and Jackson address the predicament experienced by today's scientists, who must not only collect and analyze data but also make sure their data remain useable over the life of a research program and beyond, available to readers of resulting publications as well as for potential research in the future. A recent survey confirms that researchers across the sciences are dealing with vast quantities of data (a fifth report generating data sets of 100 gigabytes or more) while at the same time lacking the resources to preserve that data sensibly (four fifths acknowledge insufficient funding for data curation).[26] Ribes and Jackson show the surprising complexities in something as apparently simple as collecting water samples from streams, while they challenge readers to think of scientists and their data as evolved and evolving symbionts, mutually dependent species adapted amid systems ecological and epistemic.

There is much more in the essays collected here than this introduction has mentioned or could encapsulate, and we hope that readers will consider as they read what the ideas are that emerge across the essays as well as what gaps there are among them. One omission, certainly, which this Introduction accentuates with its brief attention to English usage and the history of concepts, is any account of non-Western contexts or intercultural conjunctions that might illuminate and complicate data past and present. How have non-Western cultures arrived at data and allied concepts like information and objectivity? How have non-Western cultures been subject to data, in the project of colonialism, for example, or otherwise? Indeed, how are data putatively raw—and not—in non-Anglophone contexts? Do other languages deploy the food metaphor that English does? Do their speakers semantically align supposedly raw data with supposedly raw text (that is, ASCII) and supposedly raw footage (unedited film or video) the way that English speakers do? How do different languages differently resolve the dilemma of singular and plural? No collection of essays could exhaust the subject of data, of course, and that is one reason we earlier called our title a prompt rather than an argument. The authors collected in *"Raw Data" Is an Oxymoron* all hope to open the question of data, to model some of the ways of thinking about data that seem both interesting and productive, as well as to encourage further discussion. The ethics surrounding the collection and use of today's "Big Data" are a particularly pressing concern.[27]

As an additional gesture toward further discussion, we include a brief section of color images, most of them selected and described by additional contributors. The images in this color insert extend the types of data considered in this volume—some in challenging ways—while some of them also broach the important subject of representation and, more specifically, data visualization, which is not always addressed directly in the chapters that follow but which haunts them nonetheless. As the neologism "dataveillance" suggests, data provide ways to survey the world (the noun *surveillance* is related to *survey*), yet it is important to remember that surveying the world with data at some level means having data visibly before one's eyes, looking *through* the data if not always self-consciously looking *at* the data. There is then a third and final precept closely related to the other two. Not only are data abstract and aggregative, but also *data are mobilized graphically*. That is, in order to be used as part of an explanation or as a basis for argument, data typically require graphical representation and often involve a cascade of representations.[28] Any interface is a data visualization of sorts—think of how many screens you encounter every day—and so are spreadsheets, charts, diagrams, and other graphical forms. Data visualization amplifies the rhetorical function of data, since different visualizations are differently effective, well or poorly designed, and all data sets can be multiply visualized and thereby differently persuasive.

More than a few contemporary visual artists make obvious the rhetoric of data visualization: Jenny Holzer's LED feeds of poems in the place of stock quotes or headlines and "truisms" in the place of public information, for instance, confront spectators with variations on the data frames they face every day. Like the digital network, the database is an already rich and still emerging conceptual field for artwork, while a varied and variously evocative "database aesthetics" demonstrates—as we hope the chapters in this collection make clear—that recognizing the power of data visualization is an important part of living with data.[29]

Notes

1. Geoffrey C. Bowker, *Memory Practices in the Sciences* (Cambridge, MA: MIT Press, 2005), 184. This is an IBM advertising campaign from 2009 to 2010. *New York Times*, August 5, 2009, and May 13, 2011.

2. For more on data obfuscation generally, see Finn Brunton and Helen Nissenbaum, "Vernacular Resistance to Data Collection and Analysis: A Political Theory of Obfuscation," *First Monday* 16, no. 5 (May 2, 2011). The question of whether data about you is yours came before the U.S. courts in the form of a question about privacy: whether the police need a warrant to attach a

GPS device to your car and then monitor your movements. According to *United States v. Jones* (2012), they do.

3. On the bigness of data, see, for instance, Lev Manovich, "Trending: The Promises and the Challenges of Big Social Data," http://lab.softwarestudies.com/2011/04/new-article-by-lev-manovich-trending.html (accessed June 20, 2011). For an example linking big science and big data, see Peter Galison, *Image and Logic: A Material Culture of Microphysics* (Chicago: University of Chicago Press, 1997).

4. Lev Manovich, *The Language of New Media* (Cambridge, MA: MIT Press, 2001), 224.

5. See Hayden White, *Metahistory: The Historical Imagination in Nineteenth-Century Europe* (Baltimore, MD: Johns Hopkins University Press, 1975).

6. Franco Moretti, *Graphs, Maps, Trees: Abstract Models for Literary History* (London: Verso, 2005); and Michel Foucault, *The Archaeology of Knowledge & The Discourse on Language* (New York: Vintage, 1982), part 3 (French eds. 1969, 1971).

7. See Mario Biagioli, "Postdisciplinary Liaisons: Science Studies and the Humanities," *Critical Inquiry* 35 (Summer 2009): 816–833; and Mario Biagioli, ed., *The Science Studies Reader* (New York: Routledge, 1999).

8. Looking under is a gesture of "infrastructural inversion" within the sociology of knowledge; see Geoffrey C. Bowker and Susan Leigh Star, *Sorting Things Out: Classification and Its Consequences* (Cambridge, MA: MIT Press, 1999), 34–36.

9. Lorraine Daston and Peter Galison, *Objectivity* (New York: Zone Books, 2007), 51. On critique itself, see Bruno Latour, "Why Has Critique Run out of Steam? From Matters of Fact to Matters of Concern," *Critical Inquiry* 30 (Winter 2004): 225–248.

10. Joanna Picciotto, *Labors of Innocence in Early Modern England* (Cambridge, MA: Harvard University Press, 2010), 1.

11. Ibid., 2–3.

12. Daston and Galison, *Objectivity*, 27.

13. Ibid., 37.

14. Andrew Warwick, *Masters of Theory: Cambridge and the Rise of Mathematical Physics* (Chicago: University of Chicago Press, 2003), chap. 3.

15. For instance, Bill Brown, "Thing Theory," *Critical Inquiry* 28 (Autumn 2001): 1–22; Lorraine Daston, ed., *Things That Talk: Object Lessons from Art and Science* (New York: Zone Books, 2004); Lorraine Daston, ed., *Biographies of Scientific Objects* (Chicago: University of Chicago Press, 2000); Hans-Jörg Rheinberger, *Toward a History of Epistemic Things: Synthesizing Proteins in the Test Tube* (Stanford, CA: Stanford University Press, 1997).

16. Wendy Hui Kyong Chun, *Programmed Visions: Software and Memory* (Cambridge, MA: MIT Press, 2011), 11. "Thing Theory" is Bill Brown's title (see note 15).

17. Paul N. Edwards, *A Vast Machine: Computer Models, Climate Data, and the Politics of Global Warming* (Cambridge, MA: MIT Press, 2010), xiv.

18. On climate science as a form of history, see Edwards, *A Vast Machine*, xvii; on botany and bibliography, see Lorraine Daston, "Type Specimens and Scientific Memory," *Critical Inquiry* 31 (Autumn 2004): 153–182, esp. 175.

19. See Edwards, *A Vast Machine*, 17.

20. The phrase comes from a title by Theodore M. Porter, *Trust in Numbers: The Pursuit of Objectivity in Science and Public Life* (Princeton, NJ: Princeton University Press, 1995), which we recommend (along with works already cited) for readers who wish more prolonged exposure to the kinds of questions introduced here.

21. Geoffrey Nunberg, "Farewell to the Information Age," in *The Future of the Book*, ed. Geoffrey Nunberg (Berkeley: University of California Press, 1996), 117.

22. Bowker and Star, *Sorting Things Out*, 33.

23. E. F. Codd, "A Relational Model for Large Shared Data Banks," *Communications of the ACM* 13, no. 6 (June 1970): 377–387. Alan Liu led us to Codd; see his *Local Transcendence: Essays on Postmodern Historicism and the Database* (Chicago: University of Chicago Press, 2008), 239–262.

24. See Ann M. Blair, *Too Much to Know: Managing Scholarly Information Before the Modern Age* (New Haven, CT: Yale University Press, 2010); and Daniel Rosenberg, "Early Modern Information Overload," *Journal of the History of Ideas* 64 (January 2003): 1–9.

25. Nunberg, *Farewell*, 110–111.

26. See "Challenges and Opportunities," *Science* 331 (February 11, 2011): 692–693.

27. See danah boyd and Kate Crawford, "Six Provocations for Big Data," paper presented at "A Decade in Internet Time: Symposium on the Dynamics of the Internet and Society," Oxford Internet Institute, September 21, 2011; and Jay Stanley, "Eight Problems with 'Big Data,'" *ACLU.org*, April 25, 2012.

28. On mobilization and cascades, see Bruno Latour, "Drawing Things Together," *Representation in Scientific Practice*, ed. Michael Lynch and Steve Woolgar (Cambridge, MA: MIT Press, 1990), 19–68; on the effectiveness of visualizations, see, for instance, Edward Tufte, *The Visual Display of Quantitative Information*, 2nd ed. (Cheshire, CT: Graphics Press, 2001).

29. For an overview, see, for instance, Victoria Vesna, ed., *Database Aesthetics: Art in the Age of Information Overflow* (Minneapolis: University of Minnesota Press, 2007).

Color Plates

Daniel Rosenberg, Thomas Augst, Ann Fabian, Jimena Canales, Lisa Lynch, Lisa Gitelman, Paul E. Ceruzzi, Lev Manovich, Jeremy Douglass, William Huber, and Vikas Mouli

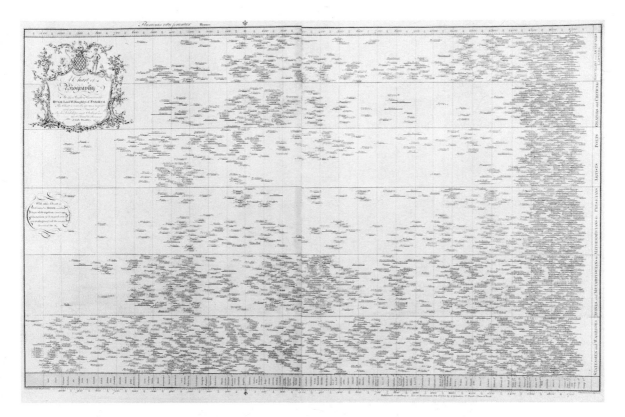

Plate 1. Chart of Biography (1765) The notion that human affairs may be studied through quantitative mechanisms was significantly advanced both in practice and theory during the seventeenth and eighteenth centuries. Theorists and philosophers from William Petty to Jeremy Bentham promoted the power of quantitative research into social and psychological phenomena, while the application of quantitative methods spread by imitation among many domains of research. Joseph Priestley's 1765 *Chart of Biography* is representative of both trends. Priestley is best remembered for his work in chemistry, including his experimental isolation of oxygen in 1772, but he made the *Chart* during his employment as a teacher at the dissenting academy in Warrington, where his duties included teaching history and politics as well as natural philosophy. In his *Chart*, Priestley applied a scientist's intuition to the problem of visualizing historical data. The *Chart of Biography* solved several problems at once. It served as an index to the dates of birth and death of famous historical figures, it clarified the relative position of these lives, and it made visible patterns of achievement in history. Priestley noted that his division of figures into categories of achievement, such as "Statesmen and Warriors" and "Mathematicians and Physicians," revealed that "the world hath never wanted competitors for empire and power, and least of all in those periods in which the sciences and the arts have been the most neglected." Priestley understood his timeline as a heuristic tool capable of making certain phenomena visible while necessarily obscuring others. Few of Priestley's immediate successors shared his reflexivity about the formal and interpretive character of the timeline. (Image courtesy of the Library Company of Philadelphia)
—Daniel Rosenberg

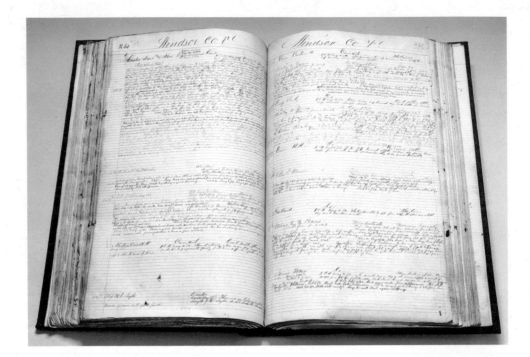

Plate 2. Moral Data   Beginning in the 1840s, commercial reporting agencies of Lewis Tappan and R. G. Dun—which together later became Dun and Bradstreet—created networks to gather and evaluate the credit of retailers throughout the United States. Detailing personal events, finances, past performance, and character traits, the companies' anonymous agents compiled histories designed to help predict the behavior of the actors in commercial markets increasingly defined by impersonality and geographic distance. Should individual merchants be trusted to perform obligations they assumed? In claiming to offer credit information that was objective and accurate, reports solicited by the agencies considered matters of property affecting the capacity to pay (bankruptcies, divorces, assets and liabilities), typical of credit reporting in our own day. They also assessed moral capacities—about honesty and the reliability of intentions, about habits of temperance, frugality, and work, in more qualitative, if not impressionist, ways, all of which counted as evidence of success and failure in nineteenth-century America. As clerks inscribed thousands of reports into large folio registers, and made them available to subscribers seeking to hedge the risks they assumed by extending credit to customers, nineteenth-century credit agencies assembled a massive, leather-bound bank of moral data that documented the evolving language in which Americans understood and diagnosed human nature in business. That language was suffused with what Scott Sandage has termed the "folklore of American capitalism," drawing on secular and religious traditions of moral cultivation, concerned with the virtues and vices, as well as Victorian regard for public opinion, for visible signs of respectability, propriety that conveyed one's "character" to strangers. Applying new technologies and systems to the storage, organization, and transmission of local knowledge, the credit agencies developed more efficient tools of information management, and in the twentieth century replaced narrative representations of financial identity with quantitative modes of fortune-telling. (*Image:* Vermont, Vol. 25, R. G. Dun & Co. Credit Report Volumes, Baker Library Historical Collections, Harvard Business School)
—Thomas Augst

WAR DEPARTMENT,
SURGEON GENERAL'S OFFICE,
ARMY MEDICAL MUSEUM,
WASHINGTON, D. C.
Photograph No.

Plate 3. "Ascertaining Capacity of Cranial Cavity by Means of Water" (1884) War Department, Surgeon General's Office, United States Medical Museum, Washington, D.C. Here is a quick lesson from the 1880s on how to measure the internal capacity of a human skull. The image captures a scene from the years when an old race science was giving way to a new physical anthropology, when the capacity of skulls might measure racial difference or offer clues to human history. Craniologists had tried beans, buckshot, and sand but worried when no two men measured alike. At the end of the century, water looked promising. Water sometimes seeped into porous bones or dripped through putty plugged into eye sockets, but sure knowledge of hydrostatics and hydraulics made water a good bet as a means to gauge a skull's capacity. You needed a collection of skulls, a beaker, a scale, and a metronome. A thin rubber lining kept water from settling into squamous sutures and leaking out through sinuses. And best of all, as one skull measurer remembered, water guaranteed objectivity, protecting scientists from the temptation to use "muscular exertion" to press a few more beans into a head. (Photography Collection, Miriam and Ira D. Wallach Division of Art, Prints, and Photographs, The New York Public Library, Astor, Lenox, and Tilden Foundations)
—Ann Fabian

Plate 4. A Bullet through an Apple (1939)   How should we interpret photographs of time intervals so small that even a flying bullet could appear perfectly immobile in midair? James R. Killian, a young science writer who would later become president of MIT and one of Eisenhower's must trusted advisors, started his illustrious career by writing about the famous strobe photographs of his colleague and friend Harold E. Edgerton. These photographs, he argued, were raw representations of the natural world. They were a "unique and literal transcription" of nature—a "scientific record" written in a "universal language for all to appreciate." Killian described Edgerton's method as a technique to "contract and expand not only space but time." His strobe was an instrument for "manipulating time as the microscope or telescope manipulates space." From Aristotle to Einstein, most scientists and philosophers felt justified in treating time as space. Although radical thinkers from Hegel to Bergson fought against this space-time conception, an orthodox interpretation of high-speed photographs as *expanding time* coalesced by mid-century. The "instantaneity" of each photograph guaranteed that these images could be studied as temporal and spatial data—easily transformed into mathematical (x, y, z, t) coordinates. But at least one anonymous observer remained skeptical, publishing a humorous critique in *The Electrical Journal* (1931). Remarking on the strobe's alleged ability to stretch time, he titled his commentary: "If Money Could Be Stretched Like That." (Copyright Harold & Esther Edgerton Foundation, 2011, courtesy of Palm Press, Inc.) —Jimena Canales

Plate 5. NORAD Santa Tracker (1964)   This image, the cover of an album titled NORAD TRACKS SANTA CLAUS, shows a military analyst peering into a radar screen at a North American Aerospace Defense Command facility, feeding data into one of the enormous off-site computers that comprised the SAGE (Semi Automatic Ground Environment) bomber detection system. It is a Cold War image with a twist, however: instead of performing customary surveillance operations, the analyst is tracking the location of Santa's sleigh on Christmas Eve. NORAD released this compilation of radio spots and assorted Christmas tunes to commemorate the ten-year anniversary of a public relations effort that began due to happenstance, but came to serve as the humorous face of an organization whose primary purpose was to monitor U.S. and Canadian airspace in anticipation of nuclear attack. In 1955, a misprint in a Sears advertisement meant that children intending to dial Sears' own Santa hot line instead reached the phone line reserved by NORAD for communication of an impending Soviet missile strikes. Playing along, the organization began to field calls from children interested in Santa's where-abouts, and eventually to issue brief broadcast "updates" of Santa's location, claiming to use NORAD's "satellites, high-powered radars, and jetfighters" to track Santa's journey. The radio broadcasts continued until 1997, when the Santa Tracker moved to the Internet. In 2007, Google partnered with NORAD on the endeavor, creating 2D Google maps and 3D Google Earth images based on NORAD's tracking data. In 2011 the Santa Tracking program drew on over 1,000 U.S. and Canadian military volunteers to field over 100,000 phone calls and emails; the Apple/Android Santa Tracker smartphone app was downloaded 1.4 million times, while the NORAD Santa Tracker Web site received 2.2 million hits. (*Image:* Bob Haynes)
—Lisa Lynch

Plate 6. Rumsfeld, Ford, and Cheney (1974)   Missing minutes of secret audio recording and other intrigues and malfeasance revealed during the Watergate scandal in the United States led to pressure for greater openness. In 1974 Congress passed a toothsome amendment to the 1966 Freedom of Information Act, but it was vetoed by President Gerald Ford. Ford vetoed the bill at the urging of his chief of staff, Donald Rumsfeld, and his deputy, Richard Cheney, who consulted with a government lawyer, Antonin Scalia ("Veto Battle 30 Years Ago," National Security Archive, nsarchive.org, November 23, 2004). Congress handily overrode Ford's veto. Responding to additional public concern about computer databases, Congress also passed the Privacy Act of 1974, which requires federal agencies to inform the public about the systems of records they use at the same time that it establishes rules for the protection of information that makes individuals identifiable. Both gestures by Congress helped to initiate the information regime in which Americans live and which has been structured further by an extended sequence of laws of fluctuating stricture and enforcement (and, in some cases, evasion) that govern privacy and the retention or destruction of records both private and public. (*Image:* Courtesy of Gerald R. Ford Library)
—Lisa Gitelman

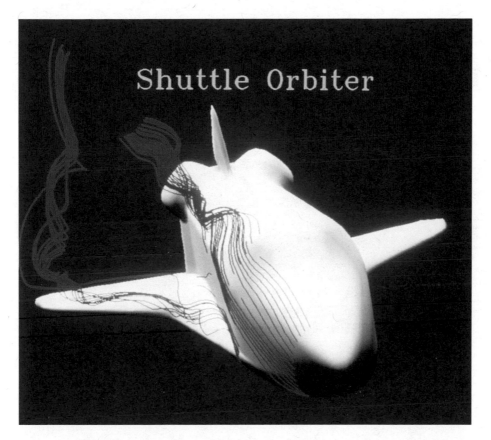

Plate 7. Aerodynamics (ca. 1980)   When the U.S. Space Shuttle was under development in the 1970s, its designers faced a number of challenges. Modeling the flow of air over the craft's wings, traditionally done using wind tunnels, was especially difficult, as the Shuttle operated over a wide range of velocities as it returned to Earth from space. Wind tunnels always involved compromises because it was impractical to replicate exactly the conditions of actual flight. Translating data obtained from a model in a tunnel into data about the flying qualities of the actual aircraft was a complex process. Shuttle designers were able to use a new tool that had just become became available: the supercomputer—a digital computer optimized for very fast numerical calculations. Usually associated with the work of computer engineer Seymour Cray, the new computers were offered first from the Control Data Corporation in the late 1960s and later on from Cray Research, a company Seymour Cray founded after leaving Control Data in the early 1970s. Supercomputers created a "virtual" wind tunnel by dividing the region around the aircraft into a grid; assigning numbers corresponding to pressure, temperature, velocity, and so on to each point on that grid; and using equations of aerodynamics to compute those values at the next step in time. To aid in analysis, these numbers were then rendered graphically in false-color, replicating the streams of smoke that were used in traditional tunnels. The image reproduced here is from the NASA Ames Research Center, in Mountain View, California, ca. 1980. It shows a supercomputer-generated image of air flowing over the right side of the Shuttle fuselage and wing. (*Credit:* NASA-Ames Research Center)
—Paul E. Ceruzzi

Plate 8. Digital Image Analysis: Manga Style Space (2010)   Images pose particular challenges for computational analysis. In 2009, we downloaded 883 Manga series containing 1,074,790 unique pages and then used our custom software system installed on a supercomputer at National Department of Energy Research Center (NERSC) to analyze their visual features, turning style into data. This visualization maps all of the pages according to their grayscale measurements, plotting the standard deviation of pixels' grayscale values in a page (x-axis) against the entropy measured over grayscale values in a page (y-axis). The pages in the bottom part of the visualization are the most graphic and have the least amount of detail. The pages in the upper right have lots of detail and texture. The pages with the highest contrast are on the right, while pages with the least contrast are on the left. Among these four extremes, we find every possible graphic variation. This suggests that our basic concept of "style" maybe not appropriate when we consider large cultural data sets. The concept assumes that we can partition a set of works into a number of discrete categories. However, the space of manga graphical variations does not have any distinct clusters, so if we try to divide this space into discrete categories, any such attempt will be arbitrary.

Plate 9. Digital Image Analysis: Manga Style Matrix (2009–2010)   We can use the same method to visualize the space of graphical variations in individual Manga series. Some series have been relatively short lived while the longest running began in 1976; the most popular Manga series contains over 10,000 pages to date. This visualization shows 192 different Manga series, each rendered as a separate scatter plot in which pages are represented by points. As in the previous visualization, the position of every point is determined by the corresponding page's visual characteristics, measured by software. (The points in the bottom part of each plot correspond to pages that are more graphic, and contain little detail, while the points in the upper right of a plot correspond to pages with lots of detail and texture.) Page order within each series is represented by color, using a blue-red gradient (pure blue—first page; pure red—last page). This mapping of page order in a series into color creates distinct visual patterns, which indicate whether visual language in a given series changes over the period of its publication. A scatter plot matrix is very useful for working with large cultural data sets. It allows us to quickly see which artifacts in a set stand out from the rest and should be investigated more closely.
—Lev Manovich, Jeremy Douglass, and William Huber

Plate 10. Visualizing the Financial Markets (2011)   First sold in 1982, the Bloomberg terminal can now be found on the desks of over 300,000 subscribers around the globe, including at investment banks, hedge funds, government agencies, and even the Vatican. A dedicated portal used to access a vast, proprietary suite of data, tools, and news, the Bloomberg terminal combines real-time quotes from a diverse array of capital and product markets—ranging from the familiar to the esoteric—with a historical repository stretching back decades, all on a single platform. As the devices have become ubiquitous in the financial sector, they have become increasingly essential for anyone hoping to understand, monitor, analyze, and participate in the modern economy. Market participants without them, or away from their desks, find themselves falling entire minutes, seconds, and nanoseconds behind. Like the telegraph and ticker before it, the Bloomberg terminal is credited with market effects. Through the process of aggregating, disseminating, and contextualizing data—modeling markets—the Bloomberg terminal actively shapes the decisions that investors make, thereby confounding the causality between principal and agent as well as human and machine. (Director of photography: Frankie DeMarco; reproduced courtesy of Roadside Attractions/Lionsgate)
—Vikas Mouli

1 Data before the Fact

Daniel Rosenberg

Is data modern? The answer depends on what one means by "data" and what one means by "modern." The concept of data specific to electronic computing is evidently an artifact of the twentieth century, but the ideas underlying it and the use of the term are much older. In English, "data" was first used in the seventeenth century. Yet it is not wrong to associate the emergence of the concept and that of modernity. The rise of the concept in the seventeenth and eighteenth centuries is tightly linked to the development of modern concepts of knowledge and argumentation. And, though these concepts long predate twentieth-century innovations in information technology, they played a crucial role in opening the conceptual space for that technology. The aim of this chapter is to sketch the early history of the concept of "data" in order to understand the way in which that space was formed.

My point of departure for this project is a happenstance textual encounter that eventually became a kind of irritation: in reading the 1788 *Lectures on History and General Policy* by the polymath natural philosopher and theologian Joseph Priestley, I stumbled on a passage in which Priestley refers to the facts of history as "data."[1] In the text, his meaning is clear enough, but the usage surprised me. I had previously associated the notion of data with the bureaucratic and statistical revolutions of the nineteenth century and the technological revolutions of the twentieth. And while I don't begrudge Priestley his use of the term, it seemed very early.

Of course, if one were to pick an eighteenth-century figure likely to be interested in data, Priestley is about as good a choice as one might make. After all, Priestley was an early innovator in the field we now call data graphics. His 1765 *Chart of Biography* is a great achievement in this field, an engraved double-folio diagram displaying the lives of about two thousand famous historical figures on a measured grid.[2] It was one of the earliest works to employ the conventions of linearity and regularity now common in

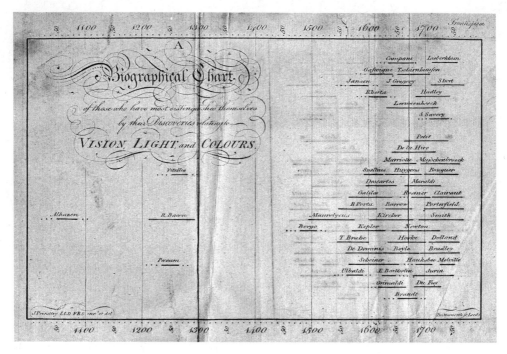

Figure 1.1   Joseph Priestley, Biographical Chart from *History and Present State of Discoveries Relating to Vision, Light, and Colours*, 1772. Biographical information extracted from *Chart of Biography* showing lives of key figures in the history of optics. Courtesy of Rare Book Division. Department of Rare Books and Special Collections, Princeton University Library.

historical timelines and the most important work of its kind published in the eighteenth century (figure 1.1).

Furthermore, Priestley was an empiricist and an experimentalist—among his many achievements, the isolation of oxygen from air in 1774 is the best remembered—and he brought an interest in aggregate phenomena to the many domains in which he researched and wrote. In his historical works, both diagrammatic and textual, Priestley was not only interested in individual facts—when was Newton born, when did he die?—but in large constellations of information. He examined fields of scientific endeavor quantitatively, grouping historical figures by their domains of achievement and plotting their lives on a measured timeline in order to observe patterns of occurrence and variations in density.

Framing historical data in a graphic such as Priestley's is second nature today, and this is in part due to Priestley himself. Today, we look at timelines and intuit historical patterns with no trouble. But all of this was new when Priestley published his charts, and the aggregate views they offered were regarded as an important and novel contribution to both social and natural science. Indeed, it is the *Chart of Biography*, not an achievement in experimental science that is named on Priestley's document of induction to the Royal Society. Later writers such as the political economist William Playfair, who debuted early versions of the line graph and bar chart in his 1786 *Commercial and Political Atlas*, credited Priestley for his innovative work in this area, too.[3]

In fact, the term "data" appears in Priestley's works many times. In his *Experiments and Observations on Different Kinds of Air*, Priestley uses "data" to refer to experimental measurements of volume. In the *Evidences of Revealed Religion*, Priestley notes that scripture offers us "no sufficient data" on the physical nature of Christ's resurrected body. In his *Essay on a Course of Liberal Education for Civil and Active Life*, Priestley writes, "Education is as much an art (founded, as all arts are, upon science) as husbandry, as architecture, or as ship-building. In all these cases we have a practical problem proposed to us, which must be performed by the help of data with which experience and observation furnish us."[4]

Nor is Priestley unique in this. The term "data" appears in a wide variety of contexts in eighteenth-century English writing. But what were these early usages? What was their importance in the language and culture of the eighteenth century? And what was their connection to the usages familiar today? What was data apart from modern concepts and systems of information? What notion of data preceded and prepared the way for our own?

All of these questions are that much more pressing since, in recent histories of science and epistemology, including foundational works by Lorraine Daston, Mary Poovey, Theodore Porter, and Ann Blair, the term "data" does heavy lifting yet is barely remarked upon.[5] Consider, for example, the first lines of Mary Poovey's landmark book, *A History of the Modern Fact*. "What are facts?" Poovey asks. "Are they incontrovertible data that simply demonstrate what is true? Or are they bits of evidence marshaled to persuade others of the theory one sets out with?" Facts may be conceived either as theory-laden or as simple and incontrovertible, Poovey says. In the latter case, we call them "data."[6]

Of course, it would not be difficult to engage in some one-upmanship. If facts can be deconstructed—if they can be shown to be theory-laden—surely data can be too. But it is not clear that such a move would be useful from either a conceptual or a practical point of view. The existing historiography of the fact is strong in its own terms, and no special harm is done by an unmarked, undeconstructed deployment of the term "data." What is more, there is a practical consideration: one has to have some language left to work with, and after thrilling conceptual histories of truth, facts, evidence, and other such terms, it is helpful to retain one or two irreducibles. Above all, it is crucial to observe that the term "data" serves a different rhetorical and conceptual function than do sister terms such as "facts" and "evidence." To put it more precisely, in contrast to these other terms, the semantic function of data is *specifically* rhetorical.

The question then is: what makes the concept of data a good candidate for something we would *not* want to deconstruct? Understanding this requires understanding what makes data different from other, closely related conceptual entities, where data came from, and how it carved out a distinctive domain within a larger conceptual and discursive sphere.

So, what was data prior to the twentieth century? And how did it acquire its pre-analytical, pre-factual status? In this, etymology is a good starting point. The word "data" comes to English from Latin. It is the plural of the Latin word *datum*, which itself is the neuter past participle of the verb *dare*, to give. A "datum" in English, then, is something given in an argument, something taken for granted. This is in contrast to "fact," which derives from the neuter past participle of the Latin verb *facere*, to do, whence we have the English word "fact," for that which was done, occurred, or exists. The etymology of "data" also contrasts with that of "evidence," from the Latin verb *vidēre*, to see. There are important distinctions here: facts are ontological, evidence is epistemological, data is rhetorical. A datum may also be a fact, just as a fact may be evidence. But, from its first vernacular formulation, the existence of a datum has been independent of any consideration of corresponding ontological truth. When a fact is proven false, it ceases to be a fact. False data is data nonetheless.

In English, "data" is a fairly recent word, though not as recent as one might guess. The earliest use of the term discovered by the *Oxford English Dictionary* occurs in a 1646 theological tract that refers to "a heap of *data*." It is notable that this first *OED* citation is to the plural, "data," rather than the singular, "datum." While "datum," too, appeared in seventeenth-century English, its usage then, as now, was limited—so limited, that in contrast to the well-accepted usage of the plural form, some critics have doubted whether the Latin *datum* was ever naturalized to English at all.[7]

"Data" did not move from Latin to English without comment. Already in the eighteenth century, stylists argued over whether the word was singular or plural, and whether a foreign word of its ilk belonged in English at all. In Latin, *data*, is always plural, but in English, even in the eighteenth century, common usage has allowed "data" to function either as a plural or as a collective singular. Guides differ, but usage authorizes both, and analogy to parallel Latin loan words gives no unambiguous guide.[8] Indeed, it seems preferable in modern English to allow context to determine whether the term should be treated as a plural or as a collective singular, since the connotations are different. When referring to individual bits or varieties of data and contrasting them among one another, it may be sensible to favor the plural as in "these data are not all equally reliable"; whereas, when referring to data as one mass, it may be better to use the singular as in "this data is reliable." According to Steven Pinker, in English today, the latter usage has become usual.[9] The fact that a standard English dictionary defines a "datum" as a "piece of information," a fragment of another linguistically complex mass noun, further strengthens this intuition.[10]

As Pinker argues, however much priggish pleasure professors may take in pointing out that the term *data* in Latin is plural, foreign plurals may be deployed in English as singulars. Were they not, we would be incorrect in referring to *an* agenda, *an* insignia, or *a* candelabra. Each of these words is a plural in its source language. Moreover, Pinker writes, "whenever pedants correct, ordinary speakers hypercorrect, so the attempt to foist 'proper' Greek and Latin plurals has bred pseudo-erudite horrors such as *axia* (more than one *axiom*), *peni*, *rhinoceri*, and . . . *octopi*." None of these exist in the source language. In the case of the last: "It should be . . . 'octopuses.' The *-us* in *octopus* is not the Latin noun ending that switches to *-i* in the plural, but the Greek *pous* (foot). The etymologically defensible *octopodes* is not an improvement."[11]

However controversial they may have been, in seventeenth-century English, neither "data" nor "datum" was particularly common. In these early years, the term "data" was still employed, especially in the realm of mathematics, where it retained the technical sense that it has in Euclid, as quantities *given* in mathematical problems, as opposed to the *quaesita*, or quantities *sought*, and in the realm of theology, where it referred to scriptural truths—whether principles or facts—that were given by God and therefore not susceptible to questioning. In the seventeenth century, in theology, one could already speak of "historical data," but "historical data" referred to precisely the sorts of information that were outside of the realm of the empirical. These were the God-given facts and principles that grounded the historian's ability to determine the *quaesita* of history.

This formulation is not marginal: technical historical practice during the early modern period involved accommodation of historical facts to scriptural data in order to make the unknown known. Some of the most heroic efforts of this sort took place in the realm of chronology, especially in efforts to correlate European and non-European historiographical traditions. Ancient records of comets and other astronomical phenomena that posed interpretive problems for histories based on scripture provide other examples. And it is notable that chronology is one of the fields in which the English word "data" flourished earliest.

In seventeenth-century philosophy and natural philosophy, just as in mathematics and theology, the term "data" functioned to identify that category of facts and principles that were, by agreement, beyond argument. In different contexts, such agreement might be based on a concept of self-evident truth, as in the case of biblical data, or on simple argumentative convenience as in the case of algebra, given $X = 3$, and so forth. The term "data" itself implied no ontological claim. In mathematics, theology, and every other realm in which the term was used, "data" was something given by the conventions of argument. Whether these conventions were factual, counter-factual, or arbitrary had no bearing on the status of givens as data.

When used in English, "data" had a much narrower meaning than did either *data* in Latin or "given" in English. Whether in mathematics, theology, or another field, use of the term "data" emphasized the argumentative context as well as the idea of problem-solving by bringing into relationship things known and things unknown. The "heap of data" that the *OED* unearthed in Henry Hammond's 1646 theological tract, *A Brief Vindication of Three Passages in the Practical Catechisme*, is not a pile of numbers but a list of theological propositions accepted as true for the sake of argument—that priests should be called to prayer, that liturgy should be rigorously followed, and so forth.[12]

It is also the case that the Latin word *data*, as a conjugation of the verb *dare*, was in constant use during the seventeenth and eighteenth centuries. In early modern Latin, as in classical Latin, *data* is everywhere. But *data* in Latin rarely translates to "data" in English. A 1733 translation of Bacon's *Novum Organum* gives a good example of the dynamic. Aphorism 105 of Book 1 of the *Novum Organum* reads as follows:

> *Inductioenim quae procedit per enumerationem simplicem res puerilis est, et precario concludit, et periculo exponitur ab instantia contradictoria, et plerumque secundum pauciora quam par est, et ex his tantummodo quae praesto sunt, pronunciat.*

> For that Induction which proceeds by simple Enumeration, is a childish thing; concludes with Uncertainty; stands exposed to Danger from contradictory Instances; and generally pronounces upon scanty Data; and such only as are ready at hand.[13]

Here we have the word "data" in the English translation, but no *data* at all in the Latin original. In fact, in the Latin, we have not even got a substantive, only the neuter substantival adjective *pauciora*, which means a small number of something—a something that Bacon's eighteenth-century translator took to be "data." All of this is made even more complicated by the fact that Bacon himself did not use the term "data" when writing in English. "Data" arrives in Bacon's corpus belatedly, posthumously, and just exactly when we would expect it, in the early 1730s.

Nor is the phenomenon of posthumous data-fication limited to Bacon. The same effect took hold in the works of Newton at virtually the same time. Bacon's translator, the physician Peter Shaw, interpolated the term "data" into the *Novum Organum* in 1733; Newton's translator, John Colson, got "data" into Newton's works three years later in 1736. In contrast to what happened in the case of Bacon, Colson did not actually put the English word "data" in Newton's mouth. But he used the term extensively in his analytic notes on Newton's works. Usually, he employed "data" in the restrictive Euclidean context in the contrast of mathematical *data* and *quaesita*. But not always. Colson's most notable usage occurs in his hagiographic introduction to his translation of Newton's *The Method of Fluxions and Infinite Series*.

> To improve Inventions already made, to carry them on, when begun, to farther perfection, is certainly a very useful and excellent Talent; but however is far inferior to the Art of Discovery, as having *pou sto* (foundations), or certain data to proceed upon and where just method, close reasoning, strict attention, and the Rules of Analogy, may do very much. But to strike out new lights, to adventure where no footsteps had been set before . . . this is the noblest Endowment that a human Mind is capable of, is reserved for the chosen few . . . and was the peculiar and distinguishing Character of our great Mathematical Philosopher.[14]

The quotation is interesting both because of the forcefulness of the distinction that it makes between the arts of invention and discovery and because of the high value that it places on the latter. Discovery, according to Colson is "the noblest Endowment" of the human mind; invention, on the other hand, is merely "useful."

From the point of view of this lexicographic history, what is most interesting is the presence of the English word "data," here used in a mode that is entirely characteristic of Colson's period. *Pou sto* is ground to stand upon, as in the famous phrase of Archimedes—"give me ground to stand upon, and I will move the world"—and undoubtedly Colson intended the phrase to be heard in this context. By the 1730s, there would have been nothing odd about using the term "data" to refer to facts discerned through experimentation, but here Colson uses "data" in the usual competing sense of principles or axioms given on the basis of which methods may be devised and facts discovered.

This is what one learns from reading. But what about the data on "data"? Might a quantitative approach be possible too? Might it be possible to study the corpus of printed English books in order to discover when "data" became a common term in English, how it was naturalized from Latin, and when it achieved its various meanings? Fortunately, today we are swimming in data for lexicographic research provided by both specialized and general databases along a spectrum from stand-alone electronic books to massive archiving and scanning endeavors such as Project Gutenberg and Google Books. Some of these resources are set up in ways that generally mimic print formats. They may offer various search features, hyperlinks, reformatting options, accessibility on multiple platforms, and so forth, but, in essence, their purpose is to deliver a readable product similar to that provided by pulp and ink. Others—still relatively few—foreground the aggregate and statistical features of the textual corpora that they access, and in a few cases

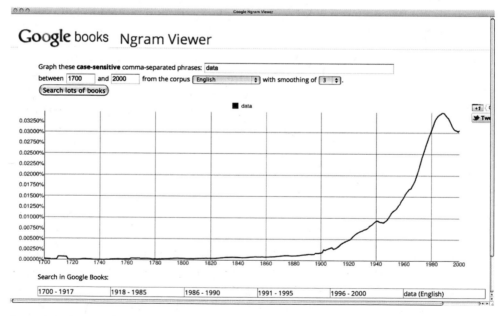

Figure 1.2   Image: Relative frequency of "data" in Google Books, by year, 1700–2000, generated by Google Ngram Viewer.

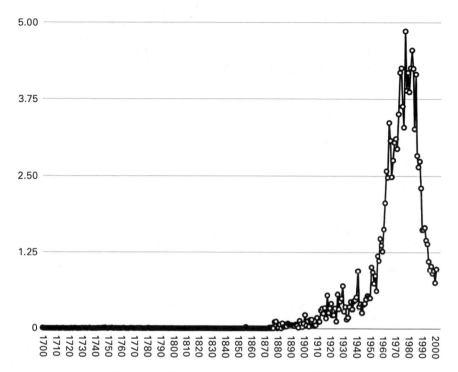

Figure 1.3   Relative frequency of "data" in Google Books corpus, 1700–2000, generated manually. *Note:* Data generated by repeated date-limited Google searches.

they do so even to the exclusion of the possibility of conventional reading, from beginning to end.

Much has been written about Google Books, but a large part of this scholarly literature has focused on the ways in which Google interacts with and places stress upon authors, publishers, libraries, and competing databases—stress that largely has to do with the fate of books in the electronic age.[15] Since the beginning of 2011, however, new attention has been focused on the research potential of Google Books as a linguistic corpus rather than as an electronic library. To facilitate research, Google has been making its book corpus accessible in two new ways: the raw data, abstracted from individual works, can be downloaded for analysis according to the interests of individual researchers, or it can be searched through a simple online interface called the Google Books Ngram Viewer. An "ngram" is a phrase consisting of a defined number of words (n): the Ngram Viewer allows corpus searches on these phrases and returns statistical results. While the Ngram Viewer is limited in the kinds of searches it can perform, its basic trick is already impressive: presented with one or more search phrases of up to five words and a historical timeframe, the Ngram Viewer can instantly produce a graph of relative usage frequency over time.

A team of Harvard researchers led by the physicist Erez Lieberman Aiden and the biologist Jean-Baptiste Michel designed the Ngram Viewer. They introduced it with a clever publicity strategy: they aimed both low and high, promoting the Ngram Viewer as both an amusing geegaw and a tool for serious scholarly research. In their January 2011 *Science* article, "Quantitative Analysis of Culture Using Millions of Digitized Books," Michel and Aiden present the Ngram Viewer as a tool for what they call *culturomics*, quantitative cultural analysis modeled on *genomics* and the other *-omic* fields booming in the natural sciences.[16]

Michel and Aiden's publicity strategy proved successful, stirring up notice in key media venues such as the *New York Times* and in the blogosphere, where the ease of use and linking prompted a lot of kitchen culturomics. Briefly, it seemed that everyone was ngramming.[17]

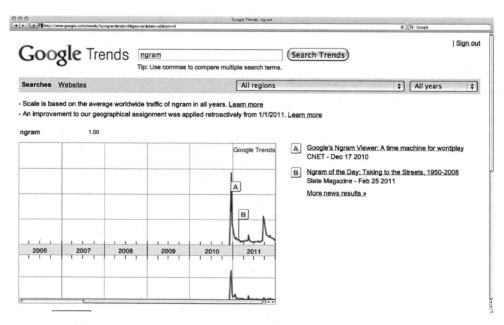

Figure 1.4   Search volume for "ngram," May 2010–December 2011, generated by Google Trends.

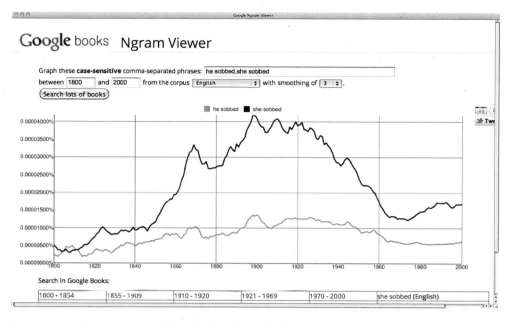

Figure 1.5 Relative frequency of "he sobbed" vs. "she sobbed" in Google Books, 1800–2000, as conceived by jezebel.com, generated by Google Ngram Viewer.

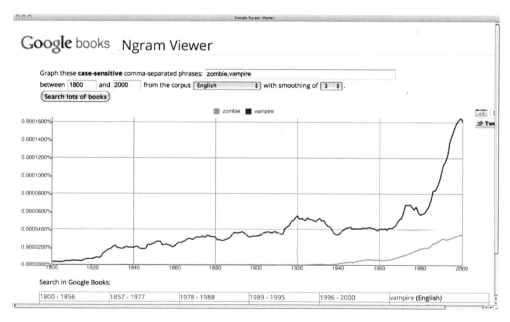

Figure 1.6 Relative frequency of "zombie" vs. "vampire" in Google Books, 1800–2000, as conceived by the-atlantic.com, generated by Google Ngram Viewer.

The Harvard team got the ball rolling with some provocative diagrams of their own, plotting the changing importance in the linguistic corpus of a variety of people, events, and things. "'Galileo,' 'Darwin,' and 'Einstein' may be well-known scientists," write Michel and Aiden, "but 'Freud' is more deeply ingrained in our collective subconscious." "In the battle of the sexes, 'women' are gaining ground on the 'men.'"[18] Even *years* themselves could be tracked through the corpus, and these produced interesting regularities.

> Just as individuals forget the past, so do societies. To quantify this effect, we reasoned that the frequency of 1-grams such as "1951" could be used to measure interest in the events of the corresponding year, and we created plots for each year between 1875 and 1975. The plots had a characteristic shape. For example, "1951" was rarely discussed until the years immediately preceding 1951. Its frequency soared in 1951, remained high for 3 years, and then underwent a rapid decay, dropping by half over the next 15 years. Finally, the plots enter a regime marked by slower forgetting: Collective memory has both a short-term and a long-term component. But there have been changes. The amplitude of the plots is rising every year: Precise dates are increasingly common. There is also a greater focus on the present. For instance, "1880" declined to half its peak value in 1912, a lag of 32 years. In contrast, "1973" declined to half its peak by 1983, a lag of only 10 years. We are forgetting our past faster with each passing year.[19]

Precisely what one makes of these word-frequency trends is, of course, open to question. "Women" are not women, nor are "men" men, and there are good bureaucratic reasons unrelated to "collective memory" why 1951 would appear in documents from 1950, but the researchers argue that within the terms of the linguistic corpus the data speaks for itself.

The value of these diagrams immediately became a subject of scholarly debate. Some humanities scholars were highly skeptical; others, such as Anthony Grafton and Geoffrey Nunberg received them more favorably. Grafton invited Michel and Aiden to address the American Historical Association in two special sessions in 2011 and 2012, the second of which was substantially devoted to rebutting misconceptions including the notion that culturomics sets out to replace historians with computer programmers.[20]

More significant than the Ngram Viewer was Google's decision to make its raw data—if the term can be applied at all—available for download so that scholars could run the numbers themselves without going through the ngram interface.[21] This resource

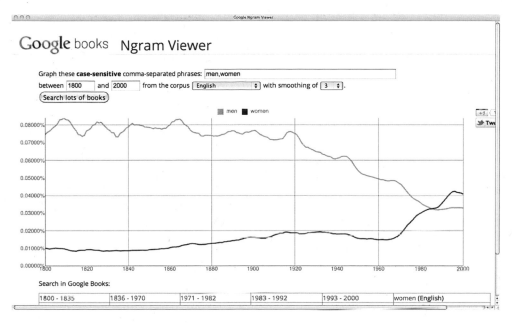

Figure 1.7   Relative frequency of "men" vs. "women" in Google Books, 1900–2000, as conceived by Michel and Aiden, generated by Google Ngram Viewer.

is likely to produce significant new research; at the same time, it should also elicit new critique.

At the time that I began research for this study, the Google Ngram Viewer was not yet available, and although it was possible to produce similar results by hand, at that time, Google Books offered neither the most obvious nor the most promising corpus with which to conduct a study such as this. As figure 1.3 demonstrates, repeating a search for the term "data" year by year and dividing the results by the results of searches for a control word in each of the same years in order to offset the effect of changing corpus size produces a curve consistent with that produced by the Ngram Viewer. This gives some indication of the promise of the corpus but only creases its surface.

In any event, I did not begin with Google Books, but rather with the subscription database Eighteenth-Century Collections Online (ECCO) from the educational publisher, Gale. ECCO is in many ways a primitive tool, and it suffers from several of the key faults for which Google Books has been criticized including inconsistent scanning quality. But ECCO has some notable advantages too. The corpus of ECCO I, based on the English Short Title Catalogue, is large, comprising more than 136,000 unique titles, 155,000 volumes, and 26 million pages of text, backed up by an accessible analog microfilm collection from which it was generated and by well-catalogued books. A later supplement, ECCO II, raises the totals to 182,000, 205,000, and 32 million, respectively. Additionally, ECCO is well defined and much more stable than Google Books, which is changing all the time. ECCO's sources are well chosen, well known, and accessible. Its out-of-the-box search functions are more flexible. And at this point in time, the metadata is much better.

In fact, there is so much that is good about ECCO that a decade ago one might have thought ECCO would have had the kind of revolutionary effect on scholarship that Google and the culturomics advocates claim Google Books will have today. ECCO has opened new research avenues, but it hasn't made that kind of impact. In 2002, ECCO's publisher promoted it as a "research revolution." A breathless review called it a "resource that scholars will die for."[22] My graduate school friends called it "the dissertation machine."

The first thing that limited ECCO's effect, of course, is that it was not made openly available like Google Books. Additionally, though ECCO is a full-text database, it does not allow users to cut and paste text. And while users can search for words under the page images, they cannot reveal what the computer sees; they cannot see the characters that the computer recognizes in the page image. Ironically, over time ECCO's publisher has loosened its rules on downloading page images. So, for database subscribers, it has become easy and quick to download page images of full books from ECCO. Yet regular users cannot even download a single page of text as interpreted by ECCO's optical character recognition (OCR) software, which suggests that over time Gale determined there is no percentage in books, not even in digitized images of books, unless the books are already packaged as data.[23]

The future is in data.

Using ECCO, I began trying to understand the sense of "data" in Priestley. Happily, my first searches turned out to be promising. On the one hand, the ECCO results are consistent with those of Google. Speaking from a strictly quantitative point of view, the big "data" takeoff is unquestionably a post-Enlightenment phenomenon. On the other hand, ECCO shows clear trends in usage in the eighteenth century that laid the foundations for all later developments, which are difficult to perceive in Google's projections. The eighteenth century produced important new ways of thinking data, and Priestley was situated, felicitously, just exactly where those new ways of thinking happened.[24]

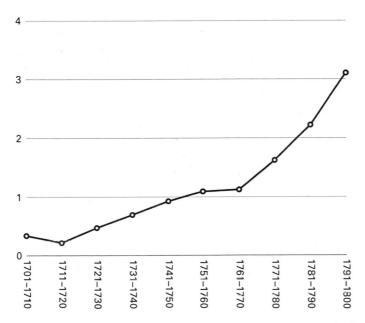

Figure 1.8   Percentage of works including the English common noun "data" in the corpus of ECCO I, 1701–1800.

The ECCO numbers are interesting, and they are also surprising in their clarity given the Google Books results, which suggest that the strong trends in the history of the term "data" begin in the nineteenth century and only accelerate definitively in the twentieth. First, from a statistical point of view, "data" was neither a rare nor an especially common term in eighteenth-century English. For comparison, a simple full-text ECCO search for the word "truth" produces hits in about 112,000 books or about 82 percent of the 136,000 total included in ECCO I. "Evidence" shows up in 66,000 books or 49 percent of total. "Fact" appears in about 35,000 or 28 percent. Even if we were to take the most generous count for "data," uncorrected for Latin usages, scanning errors, and so forth, we would find no more than 10,545 works in which "data" appears, or about 8 percent of total. And a stricter analysis of those occurrences produces a significantly smaller number, closer to 2 percent. In the eighteenth century, "data" was still a term of art.[25]

The further one goes into the data on "data," the more complicated it becomes. In my larger project, I aim to examine every usage of the term "data" in the ECCO corpus,

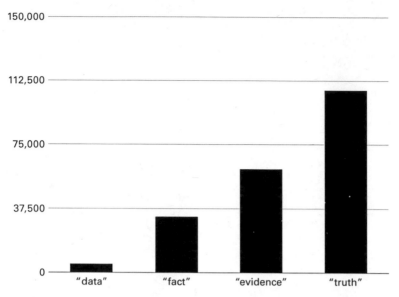

Figure 1.9   Works in the ECCO I corpus containing "data," "fact," "truth," and "evidence," 1701–1800.

not only to count for frequencies but also to examine each usage in context and to code each for semantic characteristics. The first and most pervasive problem that has turned up in this work is that a majority of usages of "data," even in the English language books in the database, turn out to be Latin. Often the Latin word *data* appears in quotations, footnotes, or conventional phrases such as *data desuper* (given from above) included in longer English texts. Other hits refer to the title of Euclid's book *Data*. Still others turn out to be scanning errors. In one instance, the search engine pulled up a reference to a certain King Data, a giant who fattened his twenty-five children by feeding them on puddings stuffed with enchanted herbs.[26] As a consequence it has been useful to examine hits individually, to sort the good from the bad and to code them, to engage in the constructive process of data making so well described in recent ethnographies of scientific practice. My own data may once have been raw, but by the time I began any serious interpretation, I had cooked it quite well.

Getting an accurate count for "data" has been a challenge. The process of scrutinizing each hit and eliminating those that were not English-language common nouns shrank the pool of viable instances. In fact, it certainly shrank the total number too far. Many works identified by ECCO as containing the word "data" in fact contain more instances

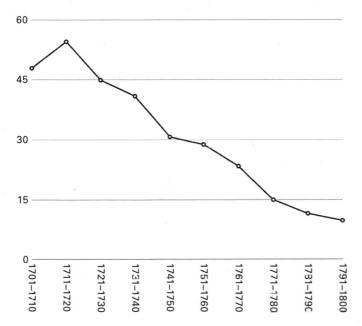

Figure 1.10   "Data" in Latin as Percentage of Total "Data" Hits in ECCO I, 1701–1800.

than ECCO shows; that is, even in works where the OCR algorithms correctly identified "data" once, they often missed it other times. And it is safe to say that there are at least as many instances in which data escaped the ECCO text search as instances in which ECCO thought it saw "data" but was mistaken. Estimating the numbers is challenging: on the one hand, there are more ways for an OCR program to overlook an instance of the word than to produce a false hit; on the other hand, since the term "data" frequently appears in a given work more than once (roughly 38 percent of the time according to my results), a significant number of OCR misses will be compensated for by correct recognitions of occurrences elsewhere in the same work.

Because the number of meaningful search hits for "data" turned out to be only about 2,300, it was possible to read them all well, to code them according to several protocols, and to produce very rich records for each instance. It was also possible to read extensively in the source works to gain a nuanced understanding of context. This has allowed me to pose a fairly wide variety of questions about the term and about key trends in its usage. And while this research is not yet complete, there are already a number of preliminary results, of which I highlight four, as follows.

First: the word "data" entered the English language in the seventeenth century and was naturalized in the eighteenth. There are a number of different sources of evidence for this, and the evidence is unambiguous. The data derived from the ECCO database shows a substantial increase in usage of the term during the eighteenth century. The number of books in which the English word "data" appears rises from 34 in the first decade of the century to 885 in the last decade, and the number of books in which "data" appears rises relative to the total number of books included in ECCO for that decade, from 0.3 percent of the total in the first decade to 3 percent of the total in the last. While this tenfold increase in relative frequency did not make data a common word, it did make it familiar. At the beginning of the century, the term "data" was italicized in the vast majority—88 percent—of cases, an indication that the word was still considered a Latin loan. By the end of the century, "data" was italicized in only 19 percent of cases. These two trends strongly reinforce one another.

Second: the term "data" came into English in the early eighteenth century principally through discussions of mathematics and theology, roughly 70 percent of instances. At

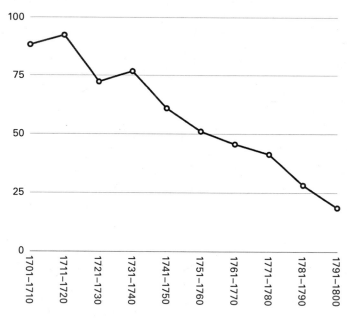

Figure 1.11   Percentage of instances of the English word "data" in ECCO I where term is italicized, 1701–1800.

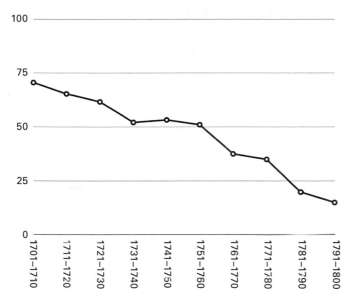

Figure 1.12   Image: Percentage of total "data" hits in English in ECCO I pertaining to Mathematics and Theology, 1701–1800.

century's end, mathematics and religion accounted for only about 20 percent of total instances, which were now dominated by empirical contexts such as those of medicine, finance, natural history, and geography.

Third: over the course of the eighteenth century, the main connotations of the term "data" shifted. At the beginning of the century, "data" was especially used to refer either to principles accepted as a basis of argument or to facts gleaned from scripture that were unavailable to questioning. By the end of the century, the term was most commonly used to refer to facts in evidence determined by experiment, experience, or collection. It had become usual to think of data as the result of an investigation rather than its premise. While this semantic inversion did not produce the twentieth-century meaning of data, it did make it possible. Still today we think of data as a premise for argument; however, our principal notion of data as information in numerical form relies on the late eighteenth-century development.

This, of course, raises an additional question. Seeing that "data" became much more commonly used during the eighteenth century, why did it take until the twentieth century for the term to become truly ubiquitous? It is clear that the fundamental semantic structure of the term "data" essential to the modern usage was settled by about 1750.

It appears, however, that while the newly outfitted term responded to and exemplified the epistemological perspective of the mid-eighteenth century, the term also was not fully required by it. Moreover, for all of the scientific achievements of the nineteenth century, the term "data" was still not of broad cultural importance. In effect, after its invention, the term went through a period of cultural latency. Though its usage expanded constantly within certain domains, throughout this period it played only a small role in the general culture. Ironically, this long period of latency may partly account for the great usefulness of the term in the twentieth century. In the twentieth century, when "data" reached its point of statistical takeoff, it was already a well-established concept, but it remained largely without connotative baggage. The arrival of computer technology and information theory gave new relevance to the base concept of data as established in the eighteenth century. At the same time, because the term was still relatively uncommon, it was adaptable to new associations.

Fourth: the *OED* is right and Google is wrong. Or at the very least, Google is not yet particularly helpful on this question. There are definitive quantifiable trends in both the currency and usage of the term "data" in the eighteenth century. It took some fairly heavy work with the ECCO data to make these trends visible, but having done it, it is clear that the *Oxford English Dictionary* account of the history of the term is mirrored in the quantitative results.

There are a number of reasons why raw Google Books results do not quite do the job for "data." First, Google Books is not yet very good or representative for periods before the nineteenth century. And even as Google Books advances, differences in the source base are still likely to pose thorny problems for quantitative comparison before the modern period. Lack of proximity search, wildcards, and other tools that aid such work as distinguishing Latin from English usages create challenges as well.

The difficulty in recognizing the true lexicographic issues in eighteenth-century English—regardless of the database one uses—is further heightened by the fact that the rise of the English-language usage of "data" during the eighteenth century coincides precisely with the decline in the general use of Latin in the Anglophone world. Without sorting, the raw numbers are highly ambiguous since the rise in the usage of "data" in English is largely offset by the decline in the use of Latin altogether. This effect is not strictly limited to the eighteenth century, but it is most significant in that transitional period.

The problem of investigating the history and semantics of "data" points to another considerable blind spot: unless search engines are full-featured, permitting good techniques of disambiguation such as proximity searching, common terms—arguably those terms we need most to understand well—may fall outside of the realm of practical investigation. Sometimes this happens by rule: for example, it is typical for search engines to exclude grammatical articles and Boolean operators from possible searches. In many cases, these restrictions virtually rule out the possibility of investigating the linguistics of conjunctions using database search functions. In other cases, blind spots of this sort are created by accident. It happens that the term "data" appears very frequently in metadata. To take one telling example: every work included in Project Gutenberg includes a legal disclaimer employing the term "data." For this reason, a simple search of Project Gutenberg to identify works in its corpus including the word "data" will produce results coextensive with the corpus itself. The online library catalog WorldCat produces another problem since it embeds the term "data" in the titles of many archival collections. None of these problems is insuperable. But none is certain to be fixed any time soon either.

It is worth adding that just because the *OED* is right and Google is wrong today doesn't mean that Google will continue to be wrong. If Google had good metadata, and if it allowed proximity searches and wildcards, we would be a long way toward being able to use it for a lot of quantitative humanities applications, whether or not one wishes to refer to these applications as "culturomic" and whether or not one regards such approaches as fundamentally new.

For the moment, it is a win for nineteenth-century reading practices, but it is not a success that is likely to stand for long. Even the venerable *OED* is moving to embrace a data-driven approach, which is as good a signal as any that we should all be ready to engage with quantitative humanities approaches in a strong, critical fashion. Among other things, as humanists, we need to pay much better attention to the epistemological implications of *search*, an entirely new and already dominant form of inquiry, a form with its own rules, and with its own notable blind spots both in design and use. In any event, I do think that my eventual results will be good news for reading even if they are not bad news for data. What is more, as we have seen with Priestley, the techniques made possible by the data-fication of our literature in many ways are consistent with ideas about ideas and writing native to the eighteenth century. In other words, at least in examining this corpus, there is a pleasing echo of the primary material, such as the charts of Priestley and Playfair, in the contemporary analytic techniques.

In the end, what does the history of the term "data" have to tell us about data today? There are a number of possible answers to this question, but one is worth particular

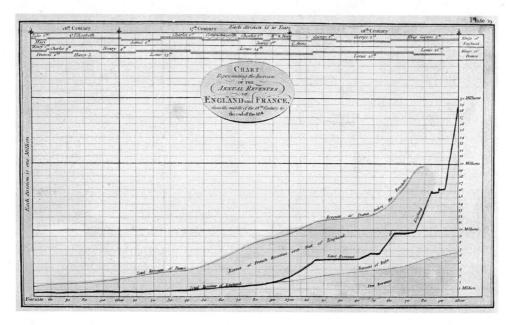

Figure 1.13   Line graph with timeline from William Playfair's *An Inquiry into the Permanent Causes of the Decline and Fall of Powerful and Wealthy Nations*, 1805. Courtesy of the Library Company of Philadelphia.

attention. This observation is supported by the numbers but not generated by them: from the beginning, data was a rhetorical concept. Data means—and has meant for a very long time—that which is given prior to argument. As a consequence, the meaning of data must always shift with argumentative strategy and context—and with the history of both. The rise of modern economics and empirical natural science created new conditions of argument and new assumptions about facts and evidence. And the histories of those terms and others in the same family nicely illustrate the larger epistemological developments.

The history of data is connected to these other histories in very important ways, but in equally important ways, it remains an outlier. Curiously, the preexisting semantic structure of the term "data" made it especially flexible in these shifting epistemological and semantic contexts. Without changing meaning, during the eighteenth century data changed connotation. It went from being reflexively associated with those things that are outside of any possible process of discovery to being the very paradigm of what one seeks through experiment and observation.

It is tempting to want to give data an essence, to define what exact kind of fact data is. But this misses the most important aspect of the term, and it obscures why the term became so useful in the mid-twentieth century. Data has no truth. Even today, when we speak of data, we make no assumptions at all about veracity. Electronic data, like the data of the early modern period, is given. It may be that the data we collect and transmit has no relation to truth or reality whatsoever beyond the reality that data helps us to construct. This fact is essential to our current usage. It was no less so in the early modern period; but in our age of communication, it is this rhetorical aspect of the term "data" that has made it indispensable.

Notes

1. Joseph Priestley, *Lectures on history, and general policy; to which is prefixed, an essay on a course of liberal education for civil and active life* (Dublin: 1788) 104. My thanks to McKenna Marsden and Dennis O'Connell for their invaluable assistance in this project.

2. Joseph Priestley, *A Chart of Biography* (London: J. Johnson, 1765).

3. Daniel Rosenberg, "Joseph Priestley and the Graphic Invention of Modern Time," *Studies in Eighteenth-Century Culture* 36 (Spring 2007): 55–104.

4. Joseph Priestley, *Experiments and Observations on Different Kinds of Air*, vol. 3 (London: 1777), 54; idem., *Discourses Relating to the Evidences of Revealed Religion*, vol. 3 (London: 1794–1799), 231; *An Essay on a Course of Liberal Education for Civil and Active Life* (London: 1765), 144.

5. Ann Blair, *Too Much to Know: Managing Scholarly Information before the Modern Age* (New Haven, CT: Yale University Press, 2010), 2; Lorraine Daston, "The Factual Sensibility," *Isis* 79, no. 3 (September 1988): 166, Theodore Porter, *The Rise of Statistical Thinking*, 1820–1900 (Princeton, NJ: Princeton University Press, 1986), 3.

6. Mary Poovey, *A History of the Modern Fact: Problems of Knowledge in the Sciences of Wealth and Society* (Chicago: University of Chicago Press 1998), 1.

7. Examples from different domains: Walter E. Myers, "A Study of Usage Items," *College Composition and Communication* 23, no. 2 (May 1972): 155–169; Carter A. Daniel and Charles C. Smith, "An Argument for Data as a Collective Singular," *ACBA Bulletin* 45, no. 3 (September 1983): 31–33; Susan E. Bates and Edward J. Benz Jr., "Troublesome Words, Linguistic Precision and Medical Oncology," *The Oncologist* 14, no. 4 (April 2009): 445–447.

8. Nearly any complete style guide will include some discussion of the problem in general or in the particular case of "data." A classic discussion of foreign loan words in English is H. W. Fowler, *The King's English*, 2nd ed. (Oxford: Clarendon Press, 1908), 26–36. On the usage of "data" in contemporary English, see *American Heritage Dictionary of the English Language*, 3rd ed.

Technical literature on the subject includes the following: Chaim Zins, "Conceptual Approaches for Defining Data, Information, and Knowledge," *Journal of the American Society for Information Science and Technology* 58, no. 4 (2007): 479–493; Carter A. Daniel and Charles C. Smith, "An Argument for *Data* as a Collective Singular," *Business Communication Quarterly* 45, no. 3 (September 1982): 31–33; Walter E. Meyers, *College Composition and Communication* 23, no. 2 (May 1972): 155–169.

9. Steven Pinker, *Words and Rules: The Ingredients of Language* (New York: Basic Books, 1999), 178.

10. Oxford Dictionaries Online, "Datum," http://oxforddictionaries.com (accessed February 10, 2012). See also Geoffrey Nunberg, "Farewell to the Information Age," in *The Future of the Book*, ed. Geoffrey Nunberg (Berkeley: University of California Press, 1996), 103–138.

11. Pinker, *Words and Rules*, 55. The eighteenth-century usage question revolved mainly around the propriety of using foreign suffixes to create plurals for naturalized loan words. "Certain it is that *effuviums* and *phenomenons* are as good as *deliriums* and *lexicons*, that therefore *effuvia* and *phenomena* are no better than *deliria* and *lexica*; that *postulata* and *data* are as great *errata* or *arcana* as *peccata* or *viscera*, *regalia* or *paraphernalia*, and (if possible) more so than *postulatum* or *datum*. . . ." James Elphinston, *The Principles of the English Language Digested*, vol. 1 (London: 1765), 6. See also, John Wesley, *Dr. Free's Edition of the Rev. Mr. John Wesley's Second Letter* (London: 1759), 5; Paul Rapin de Thoyras, *The History of England*, vol. 2, trans. N. Tindal, 4th ed. (London: 1757), 5.

12. Henry Hammond, "A Brief Vindication of Three Passages in the Practical Catechisme," in *The Workes of the Reverend and Learned Henry Hammond* (London: 1674), 248.

13. Francis Bacon, *The Philosophical Works of Francis Bacon, Baron of Verulam, Viscount St. Albans, and Lord High-Chancellor of England*, vol. 2 (London: 1733), 397.

14. Isaac Newton, *The Method of Fluxions and Infinite Series* (London: 1736), xx.

15. In this vast literature, see, for example, Nicholson Baker, *The Double Fold: Libraries and the Assault on Paper* (New York: Vintage, 2002); Robert Darnton, *The Case for Books* (New York: PublicAffairs, 2009).

16. Jean-Baptiste Michel, Yuan Kui Shen, Aviva Presser Aiden, Adrian Veres, Matthew K. Gray, The Google Books Team, Joseph P. Pickett, Dale Hoiberg, Dan Clancy, Peter Norvig, Jon Orwant, Steven Pinker, Martin A. Nowak, and Erez Lieberman Aiden, "Quantitative Analysis of Culture Using Millions of Digitized Books," *Science* 331 (2011), published online ahead of print: December 16, 2010; Erez Lieberman, Jean-Baptiste Michel, Joe Jackson, Tina Tang, and Martin Nowak, "Quantifying the Evolutionary Dynamics of Language," *Nature* 449 (2007). See also http://www.culturomics.org.

17. Patricia Cohen, "Analyzing Literature by Words and Numbers," *New York Times*, December 3, 2010; idem., "In 500 Billion Words, New Window on Culture," December 16, 2010; idem., "Five-Million-Book Google Database Gets a Workout, and a Debate, in Its First Days," December

21, 2010; idem.; Ben Zimmer, "The Future Tense," *New York Times*, February 25, 2011. Anna North, "New Google Graphs Reveal Centuries of Dicks, Pimps and Hos," December 17, 2010, http://jezebel.com/5714665/word-graphs-reveal-centuries-of-dicks-pimps-and-hos (accessed May 10, 2012); Alexis Madrigal, "Vampire vs. Zombie: Comparing Word Usage through Time," December 17, 2010, http://www.theatlantic.com/technology/archive/2010/12/vampire-vs -zombie-comparing-word-usage-through-time/68203 (accessed May 10, 2012); Dan Klein, "A Short History of Words: New Google Tool Reveals Relative Popularity of 'Shmuck,' 'Zionist,' Other Terms," December 17, 2010, http://www.tabletmag.com/scroll/53877/a-short-history -of-words/?utm_source=Tablet+Magazine+List&utm_campaign=fcdbed4176-12_20_2010 &utm_medium=email (accessed May 10, 2012).

18. Michel et al., "Quantitative Analysis," 181–182.

19. Ibid., 178–179.

20. Geoffrey Nunberg, "Counting on Google Books," *Chronicle of Higher Education*, December 16, 2010, http://chronicle.com/article/Counting-on-Google-Books/125735 (accessed May 10, 2012); Anthony Grafton, "From the President," *AHA Perspective*, March 2010, http://www .historians.org/Perspectives/issues/2011/1103/1103pre1.cfm (accessed May 10, 2012).

21. Beyond our general critique of the notion of "raw data," the specifics in this case show that the data is not raw at all. In order to correct for anomalies in the larger Google Books corpus, the Ngram Viewer, in fact, operates only on a subset of the larger data, roughly five million of the fifteen million books digitized by Google by the end of 2010. This, of course, is not a small number. Though it is only one third of the works in Google books in 2010, Michel and Aiden estimate that it represents about 4 percent of all books ever published. Michel et al., "Quantitative Analysis," 176.

22. Advertisement for Eighteenth Century Collections Online, published in *Choice* 40, no. 9 (May 2003): 1525; *Library and Information Update* 2, no. 9 (September 2003): 10.

23. Happily, users have discovered the same thing and have begun successfully pressing Gale to allow them to have limited access to scanned text and even to correct it. In April 2011, Gale authorized the Text Creation Partnership at the University of Michigan to manually key and release 2,231 texts from ECCO: http://www.lib.umich.edu/tcp/ecco/description.html (accessed May 10, 2012). Related efforts at rescanning and crowdsourced keying and correction are being organized by 18th Connect: see http://www.18thconnect.org.

24. To control for the variation in the size of the ECCO corpus from decade to decade in the eighteenth century, I divided the number of hits for "data" by the number of hits for a common control word. Experiments with several different control words suggested that using "the" as control produced a stable result.

25. The challenge of interpreting numbers such as these is highlighted by the very different results produced by a simple word search in Google compared to a parallel search in Google

Books. On the day that I composed this text, a simple search in Google for "facts" resulted in 163,000,000 hits, while "data" produced 1,160,000,000, seven times as many "data" as "facts." A Google Books search on the same day produced the inverse ratio, 166,000,000 "facts" and 28,500,000 "data," that is, six times as many "facts" as "data."

26. Antonio de Herrera y Tordesillas, *The General History of the Vast Continent and Islands of America*, vol. 3, 2nd ed. (London: 1740), 119. For more on the difficulties presented by ECCO, see Patrick Spedding, "'The New Machine': Discovering the Limits of ECCO," *Eighteenth-Century Studies* 44, no. 4 (Summer 2011): 437–453.

2 Procrustean Marxism and Subjective Rigor: Early Modern Arithmetic and Its Readers

Travis D. Williams

Mathematics is what its texts show it to be. This chapter presents an argument about how to read early modern mathematics—approached here through basic arithmetic treatises—in order to respect this maxim, and about the consequences of forgetting it. I will concentrate on the concept of rigor, perhaps the most important consideration in any attempt to read mathematics as a cultural practice, and one that necessarily encompasses all the others. I will argue that there is a reciprocal correspondence between "reading" and "rigor," so much so that to read mathematics appropriately, thoroughly, and respectfully, one must do the mathematics itself.

The relation mathematics has to its enfolding culture has everything to do with how we read or misread, or accept or reject the importance of mathematics, or even perform arithmetic correctly or incorrectly, and whether we recognize the difference. Scholarly work on early modern mathematics in its cultural context has ably demonstrated the relationship between evolving protocapitalist market economies and increasingly ubiquitous mathematical discourses related to mercantile activity, colonial expansion, newly evolving coordinations of rank, class, and land ownership, and their associated forms of power. It would be naive to assert otherwise. But this is not the whole story. The following argument will be a demonstration of this collection's core concern, that a data set is already interpreted by the fact that it is a set: some elements are privileged by inclusion, while others are denied relevance through exclusion. In the case of early modern arithmetic books, their economic function has always been included, while their other discourses have often been excluded, simply because they seem never to have been read. After a brief theoretical prologue, I will reintroduce these books. My methodology implies that, for unknown texts, description is argument; description will elicit the variety of their discourses. The second phase of my argument will show, with theoretical support, that this variety is irreducible by economic or other constraints.

The ensuing argument is both critical and metacritical; both the readers and the methodologies I will address are doubly valent, present in both our contemporary moment and in the sixteenth century. The consequence of failure is therefore a potential, in the first instance, to confuse two kinds of mathematics. The two kinds cannot be reduced to an artificial binary of old math and new math, since in that case the error would be an obvious category mistake and there would be no chance for rigor and reading to interact across time and traditions. Nor can the distinction more usefully resolve into the form it takes in recent side skirmishes of the culture wars, a contest between "good" math and the "new" math. Nor is it, in a rebarbative mode, the contradiction of opposing postures of political righteousness, notably and recently expressed by Karl Rove to a radio journalist in a discussion over the merits of different poll results prior to the 2006 U.S. Congressional election: ". . . you're entitled to your math, I'm entitled to *the* math."[1] Nor, again, is it to be reduced to a claim that the Renaissance did not always take two plus two to equal four. The Renaissance did take two plus two to equal four, though their basic arithmetic texts made more basic errors than our current arithmetic texts (even here, though, it is well not to be too smug), a divergence that is relevant to my argument. If anything, the varieties and complexities of current arithmetic, in its abstract guise of algebra, mean that two plus two can legitimately and correctly equal many things besides four more often today than at any time during the Renaissance. The most accurate description of the mathematics treated here is "ours" and "theirs," traditions that *are* related and for which reason it is especially important to recognize their divergences. My main concern then is the potential, perhaps unwitting, to confuse our reading and theirs, our rigor and theirs.

It will be useful to begin with a kind of thought experiment, as yet unencumbered by actual evidence. I posit four terms: our reading and their reading, our rigor and their rigor. Our reading is a practice of interpretation that seeks to understand the appearance and function of texts within their original historical and cultural milieus. Our reading thus incorporates the need to understand with nuance their reading: why and how contemporaneous readers would read the texts produced by their cultures. Our rigor, in relation to arithmetic, is a mathematical practice of applying well-defined procedures to arithmetic questions in order to achieve numerically correct results.[2] Their rigor might be assumed to be the same as ours, but I will hold this claim in abeyance for the moment. Immediately, we perceive a rhetorical version of a venerable arithmetic procedure, the rule of three, by which manipulation of three known terms allows us to determine a fourth term that is sought but unknown (see below for more on the rule of three). Knowing something about our reading, our rigor, and their rigor should allow us to determine something about their reading. "Something" is enough for an argument of plausibility such as this one, but that something must still be accurate. What if the three "knowns" are radically different from what we think them to be? This is where reading and rigor begin to be related. The assumption, which I now disown, that their rigor and our rigor are even similar, let alone identical, is a claim that materially informs our methodology of reading. In other terms, our assumptions about how and with what accuracy early modern readers solved arithmetic problems fundamentally affects our ability to understand how and why early modern readers read their own cultures' arithmetic texts. To put it yet another way, the assumption that our rigor was their rigor allows our reading to ignore the procedures and accuracy of their arithmetic because it can be assumed that they are the same: that their procedures are ours and that they always got the answers right, just as we assume we would do. If our reading does not do the mathematics, it is because our rigor has been assumed to be their rigor. This alters the data set and we therefore read what is not there. Reading something that is not there is the same as failing to read in a manner that is carefully and responsibly attentive to contemporary history and culture. Only if our rigor was their rigor can their reading be our reading. I will take up later what readings of ours might be mistaken for theirs, but first I will present some primary examples of early modern arithmetic in order to examine the concept of rigor, and its filiations as theirs and ours.

The Renaissance was familiar with many forms of rigor, including logic descended from classical traditions and its medieval, scholastic evolution, and the exact, accretive forms of proof associated with classical Greek geometry.[3] In the essay "Of Studies," Francis Bacon recommends mathematics "if a man's wit be wandering" in order to make them "subtle," "for in demonstrations, if his wit be called away never so little, he must begin again."[4] The Renaissance also experimented with conflations of certain knowledge and probable knowledge, syllogism and enthymeme, and logic and rhetoric in the intellectual movements we variously call humanism, Ramism, and the scientific revolution. Renaissance printing not only produced important editions of Euclid, particularly in vernacular languages, but also the "informal" geometries, by Robert Recorde, Petrus Ramus, and others, which present theorems and diagrams and informal discussion but avoid reproduction of Euclidean proofs.[5] These innovations in geometry proceeded from a relatively stable (and authoritative) tradition that had no counterpart in arithmetic. Both the form of numeration (Hindu-Arabic numbering, including its zero) and the definition of numbers and their operations (where numbers come from and what we can and cannot do with them, such as division by zero) were still new enough or unfixed enough to give arithmetic quite varied audiences, practices, and texts. The medieval distinction between "arithmetic" (often represented by Boethian theory of intervals and ratios) and "logistic" (what we today call arithmetic) broke down. The taint of manual labor associated with logistic became less serious.[6] Compared to geometry, the procedure of arithmetic is less on display, except pedagogically. Demonstration is not valorized in the same way in arithmetic; the journey to an answer is less important than the answer itself, and an achieved answer does not necessarily betray flaws in the procedure. If "clean" method (a method that is exact and theoretically grounded) and correct answers are characteristic of modern arithmetical rigor, both are less prominent in Renaissance arithmetic. Some examples of early modern arithmetic will emphasize the stakes of judging these differences with as little bias as possible.

The earliest English-language printed arithmetic treatises were designed for self-teaching. After presenting "pure," unapplied instruction in the basic operations of arithmetic, for whole numbers and fractions, these texts then present a sequence of narrative examples. What is fascinating about these examples is that their narrative specificity, the story itself, is at least as important as any mathematical content, which can often be marginal or misleading. I will address a handful of especially intriguing examples from the second edition of *An Introduction for to Learn to Reckon with the Pen*, printed in 1536 or 1537,[7] including the following:

> A drunkard drinketh a barrel of beer in the space of 14 days / and when his wife drinketh with him then they drink it out within 10 days. Now I demand in what space that his wife should drink that barrel of beer alone. For to soyle this question & such other like / ye shall first subtract the least drinker from the more / that is 10 from 14 and there remaineth 4, and that is your divisor. Now say 4 giveth 10, what giveth 14. Make it after the golden rule, and ye shall find that she should drink it in 35 days. (p8r)

The mathematical method in this example is flawed, or at least so poorly expressed that any internal logic is almost entirely absent. I will return to this later. The piquant setting and tone are reminiscent of comic fabliaux and *novelle*, including the implied expectation of moral censure on the part of the reader—the male actor is specifically labeled a drunkard—but the censure indeed is left to the reader, so as not to preclude simultaneous comic enjoyment.

> The rule and question of a testament.
> A man hath made his testament, the which hath left his wife great [with child] / & hath ordained in his testament that if she brought forth a son he should have two parts of his goods, that is to wit / of 1200 crowns. And his wife the other part / and if she brought forth a daughter, then the mother should have two parts, and the daughter the other part. It happeneth when the man is dead the wife bringeth forth a son and a daughter. I demand how shall they divide the 1200 crowns. (o2v)

The answer presented by the text is to divide the legacy into seven parts, with four parts for the son, two parts for the mother, and one part for the daughter. One must use arithmetic to make the calculations, but there is no mathematical motivation for the division into seven parts. Such a testament would probably be null and void in sixteenth-century England, and all of the parties would likely have an actionable claim for not receiving exactly his or her stated share. It is apparent that this is an attempt to maintain equity in the face of legal rigor.[8] The example also raises interesting questions about the contemporary hierarchy of genders and about the patriarchal reach of the father from beyond the grave, as we more familiarly know from the case of Portia of Belmont:

> The rule and question of the eggs.
> A young maiden beareth eggs to the market for to sell and her meeteth a young man that would play with her in so much that he overthroweth and breaketh the eggs every one, and will not pay for them. The maid doth him to be called afore the judge. The judge condemneth him to pay for the eggs / but the judge knoweth not how many eggs there were. And that he demandeth of the maid / she answereth that she is but young, and cannot well count. (o8v)

The example proceeds to explain that the girl and her mother were able to position the eggs in various geometric arrangements that provide clues to the total number. The judge is able to use this information to arrive at the answer: the young maiden started out to market with 721 eggs. The problem here is that no mathematical principle is taught that would allow one to solve this problem independently of the text. The text gives a strong hint about what the answer is likely to be and then works through some basic multiplication to verify that the hint is indeed correct. Also of interest is the touch of sexual dalliance, the mathematical ignorance of the young girl (she seems not to have studied the text in which she appears, though it is implied that the judge will have done so), the appeal to the law for redress, and the implausible, though not impossible claim that a young girl was capable of bringing sixty dozen eggs plus one to market all by herself. Other examples invoke the conventions of courtly love, chivalric romance, and biblical narrative. What are we to do with mathematics like this? There are at least two avenues of approach (and here I am being deliberately ahistoricist): (1) the mathematics and (2) everything else. I'll consider each in turn.

To make an informed conclusion about whether this mathematics is our math or their math, and if the two are really the same or not, we have to do the math, so I make this prescription: avoid the avoidance of math. We must read through the math itself, and check it. A three-fold system seems to be the minimum effort necessary to make a proper check. The first step is to solve a problem according to modern methods, if possible. This will show if the problem is capable of any mathematical solution, though we must be aware that we are situated in history as well and that our mathematical techniques and standards of rigor are as fungible as any other. The second step is to solve the problem according to the technique prescribed in the text's solution. Sometimes the technique is flawed (again, by our standards) so at this stage it is important to determine if an answer, if correct, is perhaps the result of luck. If the prescribed technique fails, then we move to the third step, which is to determine if some other method taught in the same text would solve the problem. To see how this works, let's return to the example of the drunkard and his beer. By the method generally taught for such problems today, it is immediately obvious that that the problem is soluble, and that the numerical answer provided in the text is correct.[9] Problems arise, however, when we try to apply the prescribed method of solution. The text invokes the golden rule. This is another common name for the rule of three. Here's a modern example: "If a gallon of milk costs four dollars, how much will I pay for seven gallons?" The premise posits three known amounts and one unknown amount. The rule of three provides a method to determine the unknown amount from the known amounts. Today this would be an easy problem of basic algebra: four divided by one equals x divided by seven, solve for x. The answer is $28. But since such algebraic manipulation was not yet available in the early sixteenth century, the rule of three explained how to use the grammatical arrangement of the problem to organize the data and find the solution. To return to the drunkard, the first problem to face is that the golden rule has not yet been taught under that name. It will be within a few pages, and the rule of three had been taught a few pages earlier, but for the reader learning mathematics from scratch, there is no way to know that the rule of three and the golden rule are the same thing. For such a reader, this problem is to be solved by a method that does not yet exist. Next, there is no explanation for why one should subtract the "least" drinker from the "more" (which here really means the combined rate of consumption), subtracting 10 from 14 to get 4, and no explanation for why 4 should be a divisor of anything. Lastly, the grammatical arrangement of the data, "Now say 4 giveth 10, what giveth 14," is also unmotivated by the procedure of the golden rule or the rule of three and will not produce the eventual answer. So, by the three-fold test, the problem is indeed solvable, but not by the method prescribed, nor by any other method taught in the book. For this example at least, it is safe to say that it is not our mathematics.

Now, to consider "everything else." The ostensible goal is to find a way to cut away everything that is not mathematics and thereby leave the mathematics scoured and ready for further investigation. The methodological irresponsibility of such an effort appears immediately, since the decision about what is mathematics and what is not can only proceed via a modern sense of the distinction. Some examples of current arithmetical pedagogy show that we, perhaps more than people in other eras, are particularly keen to segregate mathematics from anything that doesn't fit our sense of what mathematics should be. In an American third-grade-level mathematics textbook is a lesson entitled "Too Much Information": "At a baseball game, the Turtles scored 7 runs. The Frogs scored 6 runs and made 8 hits. How many runs were scored in the game?" During the explanation of the solution, the text tells us "The number of hits is *extra* information. Do not use 8. Find 7 + 6."[10] In a fourth-grade textbook, a lesson entitled "Too Much or Too Little Information" explains that "Sometimes when you solve problems, you may have more information than you need. Or, you may not have enough."[11] These are reasonable methods for our mathematics. The number of hits is irrelevant to the solution of the number of runs. But our mathematics does not usually care about the narrative richness of its problems, at least not explicitly. The ungendered, unracialized, anthropomorphic animals of the baseball example do tell, implicitly, a very interesting story about the role of mathematics and its instruction in American culture around the turn of the twenty-first century. Mathematicians, educators, and cultural critics will argue about whether or not these details are part of mathematics or not. They may also do so for the early modern examples quoted above. But prudence requires restraint: since their mathematics is in many ways not our mathematics, we would be unwise to cut out the narrative details and other non-numerical aspects as non-mathematical. Early modern writers and readers of mathematics were comfortable with drunkards and legal equity and courtly love in their mathematical education; we are not entitled to say they were wrong in this, or to say that these details are not mathematics, or to quietly scrub these details out of our interpretation of the tradition. This is therefore my second prescription: do not redefine their mathematics as our mathematics simply because their details are not fully readable according to our conventions.

What must be avoided for details must also be avoided for global characteristics. Each of the early modern narrative examples quoted earlier is labeled as a "rule" as well as a question, or presents a method that may be applied to "this question & such other like." These statements are an acknowledgment that similar problems may be solved by the same method. In other words, each example is a specific case of a generalized case. In current mathematics, such generalized cases are quite common and they may be expressed in a variety of forms, but especially with literal variables and algebraic notation. Some have interpreted the statement of "rules" and their associated technique to indicate that early modern readers routinely extrapolated abstracted general cases from specific examples. In a discussion of the calculation of volume from linear measurements of nonstandard Renaissance containers, Timothy J. Reiss writes of "the habit of seeing the relation between certain kinds of speculative questions and practical needs, and . . . the allied habit of seeing sure and clear mathematical rule under seeming vagary, of *discovering* universally applicable order in the real. . . . These calculations took for granted the validity of a 'systematic abstraction' and a certain spatial 'homogeneity' that assumed correspondence of some sort between 'mathematical space' and material reality: universal rule underlying perceived difference."[12] Just what this means for the nature of early modern mathematics requires the same tentative and skeptical approach as an attempt to divide narrative details from the mathematics they accompany. Given the procedural and conceptual strangeness of the narrative examples, the claim that contemporary readers would easily perceive underlying uniformity is fraught with difficulties. There is little direct evidence that they did so, except through an appeal to a zeitgeist that begins more and more to look like our own mathematics, in which general cases are both common and explicitly taught. Whatever generalization (and even the term points to its own historical unsuitability) may have existed, unrecorded, in early modern arithmetical discourses, we have no information about its form, how it was achieved, how (or even if) it was taught, and how close it was to our modes of generalization.

The preceding discussion may read as just an elaborate warning to take nothing for granted. That is not necessarily a bad thing for the cultural study of mathematics, which seems especially prone to generate unexamined assumptions in interdisciplinary historicist scholarship. The best way to avoid the imposition of narrow assumptions on the tradition of early modern arithmetic or its cultural function is to attend to the characteristics of surviving texts that are in some way alien to our own tradition. This requires us to distinguish carefully between representatives of the tradition and generalizations of it, not least because generalization is yet another concept, as we have seen, where our culture differs from theirs. It is a bold claim, but I think true nevertheless, that there is no generalized arithmetic text for sixteenth-century Britain. After a remarkable but abortive Latin-language primer, the tradition is overwhelmingly vernacular. From the 1520s, when the first English-language texts were printed, through the end of the sixteenth century, each decade wrought profound changes on the tradition. The normative form of arithmetical text shifted from monologic treatise, to humanist dialogue, and back to monologue. Sense of audience became much more refined, and there was an explosion in the diversity of texts addressed to specific applications: navigation, surveying, gunnery, fortification, commodity trading, foreign exchange, and so forth. Standards of numerical rigor and correctness changed dramatically. No text of the 1590s could stand in for any of the typical texts of the previous decades. Moreover, the books with the most enduring popularity were among the earliest in the tradition. *An Introduction for to Learn to Reckon with the Pen* went through ten editions before the middle of the seventeenth century. Robert Recorde's *The Ground of Arts*, first printed in 1543, went through forty-two editions before 1700. For both texts, each new edition brought in new material alongside the old. They became mathematical potboilers, doing a little of something for virtually every kind of reader. But as each of these innovations appeared, the older editions continued to circulate. By the 1590s and into the seventeenth century, English readers of mathematics could choose from treatises in two languages (not counting direct imports from the Continent), in several forms, from every decade since the 1520s, and with widely varying levels of crudity or sophistication, and applied to any number of specialized uses. In the face of such variety, I freely acknowledge that my examples are merely representatives from an even richer variety of texts. It is my goal to emphasize the irreducible variety of arithmetical discourses as essential to a proper understanding of what the early modern period thought it was doing when it did something it called mathematics.

Recent readers of early modern arithmetic texts, paying careful attention to their origins in the Italian *abbaco* tradition, show how mathematics was both a necessary prerequisite and natural outgrowth of early modern globalization characterized by increasing international trade and ever more complex financial instruments for credit and interest.[13] Danger, however, resides in the limitation of arithmetic's function to an economic one. The assertion of an economic teleology for mathematics may neglect a number of fruitful avenues for inquiry. In the manner of Procrustes, the mythological highwayman who would entice travelers into his house, where he would cut or stretch them to fit his bed of iron, narrowly focused approaches are in danger of mangling the tradition according to prior assumptions about the function of its texts.

"Procrustean Marxism" is the distillation of a theoretical approach to the study of early modern mathematics that is in danger of such a narrow focus. To the extent it exists at all, it is often implicit and always subtle.[14] My quarrel is not with Marxism itself. Mathematics, whether or not part of an organized educational system, is as ripe as anything to function as one of what Louis Althusser calls the Ideological State Apparatuses.[15] The danger inheres in assumptions that incorporate mathematics into vulgar conceptions of the Marxist "base" that Raymond Williams warned against.[16] More particularly, the danger is not that mathematics more properly belongs to the Marxist "superstructure" or to a base that, according to Williams, should be understood as a dynamic "process" rather than as "uniform" and "static."[17] Rather, the danger is that mathematics itself will be misunderstood as uniform and static and thus all too easily enfolded into crude models of economic imperatives. The thought processes conducive to such misunderstanding are almost proverbial: numbers don't lie; the answers to arithmetic problems are either right or wrong; the facts of arithmetic never change. Such a mathematics is immensely attractive because it provides a stability of reference that no other area of human activity can supply. Such an unchanging, unbending discipline would concomitantly be transhistorical and thus a perfect partner for the capitalism that has been rising since anyone bothered to take note of it. In an economically inflected version of the Platonic Ideal, this mathematics stands outside the contingencies of human activity. This mathematics is a collection of unquestionably objective facts and procedures, and so is easy to dismiss as "just" math, for which it would be absurd to suggest that texts, methods, and results might vary with time. This leads us back to the original, spurious assumption that our rigor was their rigor.

Williams explains that any society has a "body of practices and expectations" that form an "absolute because experienced reality beyond which it is very difficult for most members of the society to move."[18] He immediately counsels that these absolutes must be perceived in order to understand the dynamic processes of history that bring them into existence and that will eventually cause them to change into something else. The new historicism in literary and cultural studies that developed after Williams's essay and that was so influenced by it and other Marxist theory has enshrined as axiomatic the need to bracket our own contemporary absolutes to prevent them from engendering contaminating assumptions about different times and places; if such biases cannot be limited entirely, then we are obliged to be openly cognizant of them in the process of interpretation. Williams's argument is primarily addressed to the case of literature, but the conclusions he draws are clearly relevant to any discourse we might choose to study from a cultural point of view, including mathematics:

> We cannot separate literature and art from other kinds of social practice, in such a way as to make them subject to quite special and distinct laws. They might have quite specific features as practices, but they cannot be separated from the general social process. . . . When we read much literature, over the whole range, without the sleight-of-hand of calling Literature only that which we have already selected as embodying certain meanings and values at a certain scale of intensity, we are bound to recognize that . . . . this activity takes place in all areas of the culture.[19]

Since the purpose of resurrecting early modern arithmetic might be to provide it with an instrumental role in economic systems that *are* historically contingent, it would be highly dubious to assume that mathematics itself is untouched by history. Williams argues that though we may be drawn to the "irreducible individuality [of works], we should find ourselves attending first to the reality of their practice and the conditions of the practice as it was then executed."[20] He continues to emphasize that this will require us to ask different kinds of questions, but the larger point is that we must remember to ask both kinds of questions, and not allow a bias toward one kind of question and its answers to lead to neglect of the other. In analogy with literature that is unwarrantably confined to "Literature," mathematics may suffer the similar fate of becoming "Mathematics." There are several "meanings and values" of Mathematics—which also happen to be recognizable as absolutes of our own culture—to guard against. First, there is the truth that contemporary mathematics, both simple and advanced, underpins every area of contemporary economic activity. It does not follow that mathematics was developed, either in the early modern period or in any other, solely to promote our current economic systems. Second, the contemporary stereotype that mathematics and its practitioners are cold, impersonal, remote, and unsociable could be disproved for our time and for many others.[21] That such characteristics are organically related to the role mathematics may or may not have played in the rise of "depersonalized" economic systems is equally dubious.

The arithmetic of early modern Britain is not our arithmetic. Its preoccupations, variety, and inscrutability render it distinct. Some of its practices today could only be described as errors and swept aside. But a responsible historiography must make its task to read that error as significant, even as potentially central to whatever early modern arithmetic was doing for its own readers in its own culture. To ignore these qualities and focus instead on those aspects that are similar to our own mathematics is to force a teleology that is not there. It is unhelpful to call early modern arithmetic the "early modern universal,"[22] not because early modern arithmetic was not doing substantial things for its culture, but because the epithet limits that function to one that is similar to what our arithmetic does for us, today.

The earliest English arithmetic primer is a Latin-language text that deploys humanist eloquence as a major focus of mathematical discourse, a focus that is deliberately posited as a refined opposition to comparatively crude instrumental techniques of finance and exchange. The arithmetic book written by Cuthbert Tunstall, sometime bishop of London and Durham, is the *De arte supputandi libri quattuor* of 1522. "Supputation" means reckoning or accounting, and *Four books on the art of supputation* presents all the usual mathematical skills a merchant or trader would need to keep proper accounts. Tunstall dedicates his work to his close friend Sir Thomas More and recommends the book to More's children, "for by nothing are the abilities of young folk more invigorated than by the study of mathematics."[23] He expresses particular concern that such treatises as he had already consulted are so vilely expressed that he takes it as his duty to cast arithmetic into Latin as eloquent as such a barbarous subject can sustain. Though Tunstall claims, in a familiar pose of humanist deprecation, that he "did not consider [his work] worthy to come into the hands of learned men," he nonetheless hoped it would approach some kind of eloquence: " . . . many points often arose which seemed to offer no scope either for Latin style or for eloquence, . . . More than once I reflected that, even though I were not able to realize to the full my hope that everything should glitter with more or less brilliance, it would not be without use to render some matters less uncouth which, in their rude state, lay neglected." Tunstall is also clear on another of his purposes. He undertakes the task of writing a book so far out of his own professional domain precisely because he feels he has often been cheated by merchants and money changers and believes that only independent knowledge of arithmetic can prevent one from being cheated: to "avoid the trickery I greatly suspected" in dealings with money-lenders. Whatever a reader might do with Tunstall's book, to whatever use the mathematics in it might be put, the author's own goal is *caveat emptor*, a lesson moreover couched in a Latin style that suggests that a mathematics book is not just good for the technique a reader can get out of it, but also for the aesthetic engagement of actually reading it. At the very start of the English printed mathematics tradition, we confront the possibility that a math book might be a good read, something that need have no essential connection to the economic uses of the mathematics contained therein.[24] Additionally, in Tunstall's world, individual mathematical knowledge, itself couched in the eloquence prized by humanist thought, is at least as important and potent a commodity as anything actually traded or paid for.

Tunstall's work does not seem to have been very influential in England; after the original edition of 1522 (London), all subsequent editions were published in Paris (1529, 1535, 1538) or Strasburg (1543, 1544, 1548, 1551).[25] When his book left England so, for the most part, did Latin-language arithmetic instruction. The subsequent tradition is almost entirely vernacular.[26] Within a decade of Tunstall's first edition, the dominant form of English-language arithmetic was the self-teaching primer containing the narrative examples discussed earlier.

On the heels of the narrative tradition in mathematics, Robert Recorde developed his extensive curriculum of mathematical subjects, mostly cast as humanist dialogues. His extant books cover arithmetic, geometry, astronomy, and algebra.[27] Along with mathematics, Recorde deals extensively with the concerns of sixteenth-century humanists: vernacular learning, civic responsibility, textual editing, and the reform of the curriculum. Amid this variety of uses to which mathematics can be applied, Recorde of course includes those of merchants. Mercantile examples appear frequently in most of the books, but Recorde seeks to distance himself from the task of writing a merely utilitarian merchant's arithmetic. He expresses some disdain for the "vulgar" covetousness of merchants, as in this passage from the algebra book, the first in English, entitled *The Whetstone of Wit* (1557): "Many praise [mathematics], but few do greatly practice it: unless it be for the vulgar practice, concerning Merchants trade."[28] Indeed, in his astronomy text of a year before, *The Castle of Knowledge*, Recorde priggishly accepts only one kind of covetousness: "knowledge [may] well be compared to covetousness: for as the covetous mind with getting is never satisfied, so knowledge by knowing doth covet still more."[29] In 1552 he reworked and expanded his 1543 *Ground of Arts*, a text that itself obliquely seeks to challenge the role of grammar as the ancient foundation of the liberal arts. The new edition is dedicated to King Edward VI and incorporates a lengthy dedicatory preface that emphasizes the importance of mathematics to the smooth operation of the commonwealth, and a necessary skill for every member of it, from king to plowman. During a discussion of surveying in the text proper, the conversation between the Master and Scholar becomes topical, touching on the contentious issue of enclosures and common tillage. In this exchange, the Master and Scholar have determined how many sheep may be kept on 7,000 acres:

Scholar   I see by this rate he that hath 7000 acres of ground may keep 20,000 sheep, & thereby I conjecture that many men may keep so many sheep. For many men (as the common saying is) have so many acres of ground.

Master   That talk is not likely, for so much ground is in compass above 48 3/4 miles. but leave this talk, & return to your questions, lest your pointing be scarce well taken.

Scholar   Indeed I do remember that the Egyptians did grudge so much against shepherds, till at length they smarted for it, & yet they were but small sheepmasters to some men that be now, and the sheep are waxen so fierce now and so mighty, that none can withstand them but the lion.

Master   I perceive you talk as you hear some other [do], but to the work of your question.[30]

These fierce sheep, often known as Cotswold lions, probably originated in Sir Thomas More's *Utopia* (1516): "Your sheep that were wont to be so meek and tame and so small eaters, now, as I hear say, be become so great devourers and so wild, that they eat up and swallow down the very men themselves. They consume, destroy, and devour whole fields, houses, and cities.[31] In More's text, the conversation develops into an extended discussion of the evils of enclosure. In Recorde's text, the dialogue form provides the opportunity for effective political commentary within the safety of the indirect exchange. It is the young scholar who makes the most piercing comments, while it is the experienced master who warns him about being too piercing, and then urges him to get back to the mathematical matter at hand. As in Tunstall's preface, the economic stakes are obvious, but the commentary has a richness that defies reduction to a characterless instrument of market forces. The stakes are indeed ratcheted up with the comparison of the contemporary English controversies over enclosure with Egyptian problems with shepherds, a clumsy and inapt allusion to the Hebrew enslavement by Pharaoh. It is not clear if the scholar means to align the Egyptians or the Hebrews with the shepherds. (In More, it is clear that upper-rank enclosers are the villains, hurting lower-rank arable farmers.) And if the English farmers, pushed off their land by enclosures, are the aggrieved party, how they are meant to align with either the Egyptians or the Hebrews is also not clear. In any case, the scholar recovers from this moment to obliquely praise the king. Recorde, through the scholar, implies that the real lion is not the Cotswold sort, but the king himself, who with the aid of mathematical sciences can reform these economic problems. State power thus may use mathematics to regulate the excesses of economically rapacious enclosers.

What mathematics was and did in early modern Britain may only be understood from the archival record. It certainly contributed to incredible economic growth that we rightly recognize as a forerunner of our market economies. Its limitations perhaps also led to behaviors that exacerbated the inflation that occurred in the sixteenth and seventeenth centuries. It also did a variety of other things with no necessary relationship to buying and selling. From our point of view, early modern British arithmetic is clumsy and stunted, but it nevertheless worked for its own time and place. Its practices are often transparent to us, but just as often they are opaque. We must approach early modern mathematics provisionally, and be ready to question our own methodologies and proclivities with explicit rigor. I do not exempt myself: this argument's limitation to basic arithmetic itself deserves scrutiny, not least because the highly developed and rigorous tradition of geometry was concurrently altering the management and commodification of land, and the economic alienation of those who worked on it. A restructured historiography of early modern arithmetic may complicate its participation in evolving economies with a fascinating diversity of motives. We can only begin to name these motives and interpret their larger significance if we read early modern mathematics without presuming to know exactly what "mathematics" should mean. Such an undertaking is daunting, but one way to find our footing through this dark maze is to do the mathematics as an integral part of reading it.

Notes

1. "Rove Sees No GOP Fall in the 2006 Election," *All Things Considered*, National Public Radio, October 24, 2006. The italics are my addition, an interpretation of Rove's tone in the online archive recording of the radio interview at http://www.npr.org (accessed August 20, 2010).

2. "Rigor" as I use it here has a long historical pedigree that takes in traditions of methodological strictness, necessity, and Euclidean demonstration. It also has relation to objectivity, particularly "structural objectivity" as investigated by Lorraine Daston and Peter Galison: "the aspects of scientific knowledge that survive translation, transmission, theory change, and differences among thinking beings due to physiology, psychology, history, culture, language, and . . . species"; *Objectivity* (New York: Zone Books, 2007), 256. Daston and Galison show that, of course, such solidity is a chimera.

3. See Peter Dear, *Discipline and Experience: The Mathematical Way in the Scientific Revolution* (Chicago: University of Chicago Press, 1995), for discussion of early modern correlations of Euclidean method with Aristotelian logic.

4. Francis Bacon, *The Philosophical Works of Francis Bacon*, ed. John M. Robertson (London: Routledge, 1905), 798. In *De Augmentis Scientiarum* (book 6, chapter 4) Bacon makes a similar point: "if one be bird-witted, that is, easily distracted and unable to keep his attention as long as he should, Mathematics provides a remedy; for in them if the mind be caught away but a moment, the demonstration has to be commenced anew" (*Philosophical Works*, 560).

5. In his geometry, *The Pathway to Knowledge* (London: Reynold Wolfe, 1551), Recorde constantly promised to provide the formal proofs at a later time. He never got around to doing them. Ramus and his followers deliberately avoided the formal proofs as not conducive to effective pedagogy. See Michael S. Mahoney, "The Beginnings of Algebraic Thought in the Seventeenth Century," in Descartes: *Philosophy, Mathematics and Physics*, ed. Stephen Gaukroger (Sussex, UK: Harvester, 1980), 141–155: "What particularly displeased Ramus about Greek mathematics as it stood in the transmitted texts was precisely its rigour, which he took to be poor method" (149).

6. In the opening exchange of Recorde's arithmetic text, *The Ground of Arts*, the Scholar only grudgingly agrees to study mathematics, which he slights as common: "Sir, such is your authority in mine estimation, that I am content to consent to your saying, and to receive it as truth, though I see none other reason that doth lead me thereunto: whereas else in mine own conceit it appeareth but vain, to bestow any time privately in learning of this thing, that every child may, and doth learn at all times and hours, when he doth any thing himself, and much more when he talketh or reasoneth with others." The Master rebuts by arguing that the commonness of mathematics makes it essential knowledge for everyone; *The Ground of Arts* (London: Reyner Wolfe, 1543), A2v.

7. This is the second edition of a series of texts that went to ten editions by 1629. The 1536–1537 edition is the first of which we possess a complete copy. The first edition, from 1526 (London: Richard Fakes), survives only as a single leaf, now held in the British Library. I have modernized spelling in transcriptions of sixteenth-century texts. Parenthetical references are to signatures.

8. For historical background in mathematics applied to the law, see A. I. Sabra, "The Appropriation and Subsequent Naturalization of Greek Science in Medieval Islam: A Preliminary Statement," *History of Science* 25 (1987): 223–243; David Eugene Smith, "On the Origin of Certain Typical Problems," *American Mathematical Monthly* 24 (1917): 64–71.

9. A modern solution would make liberal use of modular algebraic notation. If the husband drinks the entire barrel in 14 days, then he drinks 1/14 of the whole in one day. The husband and wife together drink the whole in 10 days, so they drink 1/10 of the whole in one day. The

wife drinks the barrel in an unknown number of days, "w," so she drinks 1/w in one day. Therefore 1/14 + 1/w = 1/10; solve for w, which turns out to be 35 days.

10. L. Carey Bolster et al., *Invitation to Mathematics* (Glenview, IL: Scott, Foresman, 1987), 82–83.

11. Randall I. Charles et al., *Scott Foresman-AddisonWesley Math* (Menlo Park, CA: Scott Foresman-Addison Wesley, 1998), 186–187.

12. Timothy J. Reiss, *Knowledge, Discovery and Imagination in Early Modern Europe: The Rise of Aesthetic Rationalism* (Cambridge, UK: Cambridge University Press, 1997), 143–144. Reiss quotes terminology from Erwin Panofsky, *Perspective as Symbolic Form*, trans. Christopher S. Wood (New York: Zone, 1991), originally published as "Die Perspektive als 'symbolische Form,'" *Vorträge der Bibliothek Warburg* 4 (1924–1925): 258–330. Panofsky's totalizing argument about perspective has been the subject of extensive critique, notably in James Elkins, *The Poetics of Perspective* (Ithaca, NY: Cornell University Press, 1994). The calculation of the volumes of nonstandard containers was a widespread application of mathematics in early modern Europe undertaken by specialists called "gagers" and for which there was a specialty mathematical literature. I only object to the imposition of a culture of abstraction and generality when there is no evidence for it. For more on gaging, see David Eugene Smith, *History of Mathematics*, vol. 2 (New York: Ginn, 1923–1925), 580–581.

13. "Recent" is an elastic term that takes in the last forty years. There are also some older studies, from the beginning and middle of the twentieth century, that still deserve attention today. See Denise Albanese, "Mathematics as Social Formation: Mapping the Early Modern Universal," in *The Culture of Capital: Property, Cities, and Knowledge in Early Modern England*, ed. Henry S. Turner (New York: Routledge, 2002), 255–273; Paula Blank, "Shakespeare's Equalities: Checking the Math of *King Lear*," *Exemplaria* 15 (2003): 473–508; Natalie Zemon Davis, "Sixteenth-Century French Arithmetics on the Business Life," *Journal of the History of Ideas* 21 (1960): 18–48; Richard Goldthwaite, "Schools and Teachers of Commercial Arithmetic in Renaissance Florence," *Journal of European Economic History* 1 (1972): 418–433; Kenneth J. Knoespel, "The Narrative Matter of Mathematics: John Dee's Preface to the *Elements of Euclid of Megara* (1570)," *Philological Quarterly* 66 (1987): 27–46; Samuel Lilley, "Robert Recorde and the Idea of Progress. A Hypothesis and Verification," *Renaissance and Modern Studies* 2 (1958): 3–37; Carla Mazzio, "The Three-Dimensional Self: Geometry, Melancholy, Drama," in *Arts of Calculation: Quantifying Thought in Early Modern Europe*, ed. David Glimp and Michelle R. Warren (New York: Palgrave Macmillan, 2004), 39–65; Eugene Ostashevsky, "Crooked Figures: Zero and Hindu-Arabic Notation in Shakespeare's *Henry V*," in *Arts of Calculation: Quantifying Thought in Early Modern Europe*, ed. David Glimp and Michelle R. Warren (New York: Palgrave Macmillan, 2004), 205–228; Shankar Raman, "Death by Numbers: Counting and Accounting in *The Winter's Tale*," in *Alternative Shakespeares 3*, ed. Diana E. Henderson (New York: Routledge, 2007), 158–180; Reiss, *Knowledge, Discovery and Imagination*; Smith, "On the Origin of Certain Typical

Problems"; Frank Swetz, *Capitalism and Arithmetic: The New Math of the 15th Century* (La Salle, IL: Open Court, 1987); Frank Swetz, "Fifteenth and Sixteenth Century Arithmetic Texts: What Can We Learn from Them?" *Science and Education* 1 (1992): 365–378; Henry S. Turner, *The English Renaissance Stage: Geometry, Poetics, and the Practical Arts 1580–1630* (Oxford: Oxford University Press, 2006); Linda Woodbridge, ed., *Money and the Age of Shakespeare: Essays in New Economic Criticism* (New York: Palgrave Macmillan, 2003).

14. See, for example, Raman, "Death by Numbers": "But [Hindu-Arabic notation's] spread [in Europe] and popularisation required the further impetus of commercial expansion, as well as the displacement of the teaching of commercial arithmetic from the university curriculum into the mercantile world. These conditions were met in late fifteenth-century northern Italy, where the growth in the complexity of commercial transactions and trade gradually led to widespread adoption of the new notation" (179n14); Turner, *English Renaissance Stage*: "The geometrical manuals were sponsored also by the protracted colonial expansion of the Elizabethan state into Ireland, and by its commercial and political interests overseas in the New World or across the Channel on the Continent" (33); Albanese, "Mathematics as Social Formation": "As harbinger of a social formation associated with the market as well as with shifting forms of knowledge production, Tudor mathematics thus stands as a structure of feeling for early modernity" (258). Albanese's essay is remarkable for being both an astute consideration of the status of early modern mathematics as contingent and unfixed in its relationship to other modes of social formation, and a study that should be approached carefully with respect to the details it records about the texts and writers of the tradition, particularly the nature of Robert Recorde's humanist project.

15. Louis Althusser, "Ideology and Ideological State Apparatuses (Notes towards an Investigation)," in *Lenin and Philosophy and Other Essays*, trans. Ben Brewster (London: NLB, 1971), 121–173. Althusser mentions mathematics several times in this regard (127, 147).

16. Raymond Williams, "Base and Superstructure in Marxist Cultural Theory," *New Left Review*, 1st ser., 82 (1973): 3–16. My use of Althusser's and Williams's widely known essays is methodologically deliberate. These essays have entered the canon of theory pedagogy and are known to all literary scholars and cultural historians, even those who are not avowed specialists in Marxist theory. The essays correlate appropriately with the casual, implicit recourse to economic explanations of mathematical activity that are my concern.

17. Ibid., 5.

18. Ibid., 9.

19. Ibid., 13.

20. Ibid., 16.

21. See Amir Alexander, *Duel at Dawn: Heroes, Martyrs, and the Rise of Modern Mathematics* (Cambridge, MA: Harvard University Press, 2010); Douglas M. Jesseph, *Squaring the Circle: The War*

between Hobbes and Wallis (Chicago: University of Chicago Press, 1999); Matthew L. Jones, *The Good Life in the Scientific Revolution: Descartes, Pascal, Leibniz, and the Cultivation of Virtue* (Chicago: University of Chicago Press, 2006); Helena M. Pycior, *Symbols, Impossible Numbers, and Geometric Entanglements: British Algebra through the Commentaries on Newton's Universal Arithmetick* (Cambridge, UK: Cambridge University Press, 1997).

22. Albanese, "Social Formation." Elsewhere, Albanese calls early modern mathematics "a new cultural dominant" (259).

23. Cuthbert Tunstall, *De arte supputandi libri quattuor* (London: Richard Pynson, 1522); English translations from Charles Sturge, *Cuthbert Tunstall: Churchman, Scholar, Statesman, Administrator* (London: Longmans, Green, 1938), 71–78.

24. It is also true, though, that a book worth reading is itself a valuable commodity that may be produced, priced, marketed, and sold with the aid of the mathematical techniques taught therein.

25. On the editions of *De arte supputandi*, see Sturge's Appendix XXVI.

26. Contra Albanese, "Social Formation," 263–264. Tunstall's text has never been thoroughly examined. There is important work to be done on it, including how it may preserve for print culture the preceding, and largely lost, manuscript tradition of mathematics instruction.

27. A fifth text authored by Recorde has also survived, a traditional Galenic treatise on uroscopy: *The Urinal of Physick* (London: Reyner Wolfe, 1547).

28. Robert Recorde, *The Whetstone of Wit* (London: John Kingston, 1557), B3r.

29. Robert Recorde, *The Castle of Knowledge* (London: Reyner Wolfe, 1556), A1r.

30. Robert Recorde, *The Ground of Artes* (London: Reyner Wolfe, 1552), Y7v-8r. I have expanded speech headings.

31. This is Ralph Robinson's translation from 1551, in Thomas More, *Utopia*, ed. Richard Marius (London: Dent, 1994), 26. More's original Latin: "Oues inquam uestrae, quae tam mites esse, tamque, exigeo solent alo, nunc (uti fertur) tam edaces atque indomitae esse coeperunt, ut homines deuorent ipsos, agros, domos, oppida uastent ac depopulentur"; Thomas More, *The Complete Works of St. Thomas More*, vol. 4, ed. Edward Surtz and J. H. Hexter (New Haven, CT: Yale University Press, 1963), 64–66.

3

From Measuring Desire to Quantifying Expectations: A Late Nineteenth-Century Effort to Marry Economic Theory and Data

Kevin R. Brine and Mary Poovey

In 1891, an ambitious young doctoral candidate in New Haven drew up a plan for an elaborate mechanism, whose operations were intended to help readers visualize how the economy worked.[1] Two years later, with financial help from a colleague, the same young man, now an assistant professor of mathematics at Yale, turned his plan into a three-dimensional machine, which demonstrated in real time and space the economic principles he had described in his dissertation. Three years after that, in 1896, our mathematician-turned-economist undertook two additional projects: he tried to stabilize the definition of the term "capital"; and he attempted to put theory in a quantitative form and test his theoretical hypothesis against the available data. In this series of moves—from a drawing to a three-dimensional machine to a quantitative formulation that could use and be tested against empirical data—Irving Fisher simultaneously helped liberate the academic discipline of economics from its nineteenth-century polemical phase and created the prototype for what is now the normative way that economists make truth claims about the economy.[2] In the process he exposed something peculiar about the nature of the data economic claims invoke: the numbers that seem simply to represent actual economic events are actually the products of a complex historical and practical process that has made them useful *for the formulation in which they appear*. This process, moreover, embeds aspects of the assumptions most economists now take for granted into the data themselves. Not only is economic data never raw, then, in the sense of being uninterpreted, but also the form that makes data suitable for economists' use carries with it assumptions about quantification and value that now go unnoticed and unremarked.

In this chapter we argue that the representational journey Irving Fisher took during this five-year period—from drawing to machine to quantitative formulation—culminated in a series of breakthroughs that helped recast modern economic knowledge.

What initially looks like nothing more than a sequence of representational alternatives, in other words, contributed to a wholesale revision of the discipline, which turned on the new importance economists assigned to data *and* the peculiar nature of the data that a revised economic methodology could use. To see how Fisher reimagined economic knowledge and why the data modern economists use might be considered peculiar, we need to follow in some detail Fisher's brief, but consequential, representational journey.

From Mechanics to Quantification: The Analogy Machine

The economic theory that Fisher's dissertation was designed to illustrate involves value—or, more specifically, the way that desire, which most late nineteenth-century economists assumed to motivate supply and demand, affected prices and the quantities of commodities bought and sold.[3] In that period, as now, this theory was known as equilibrium. Essentially, it stated that, under idealized conditions, the prices that economic markets establish reflect an exchange equilibrium, or balance, between the amount of buyers' desire for goods or services and the prices sellers want for those goods or services. In the language of the day, desire was generally expressed in terms of "utility," as Fisher's description of his topic reveals, and the last "point" at which a unit of any given commodity was still considered desirable was expressed in terms of its "marginal utility." What he wanted to illustrate, Fisher explained, was "the dependence of value on utility, disutility, and commodity, the equality of utilities, the ratio of utilities, the utility of a commodity as a function of the quantity of that commodity solely, or of that commodity and others, conjointly."[4] In his prefatory remarks, Fisher also surveys the kinds of analysis that might clarify these relationships:

> How few scholars of the literary and historical type retain from their study of mechanics an adequate notion of force! Muscular experience supplies a concrete and practical conception but gives no inkling of the complicated dependence on space, time, and mass. Only patient mathematical analysis can do that. This natural aversion to elaborate and intricate analysis exists in Economics and especially in the theory of value. The very foundations of the subject require new analysis and definition.[5]

Alluding to the purely verbal method favored by late nineteenth-century political economists, Fisher dismisses "literary and historical" analysis; then he dismisses mere "muscular experience," because it cannot factor in *as abstractions* the Euclidean conditions in which economic relations must be conceptualized ("space, time, and mass"). "Patient mathematical analysis" emerges as the only method capable of giving an "inkling" of the complicated relationships that constitute value. Even though Fisher had discovered a method that he thought capable of illustrating the problem of value, however, he still needed to figure out how to identify the quantities, especially of demand or desire, to which to apply the math.

In *Mathematical Investigations*, Fisher tried to solve this problem with an elaborate "mechanism," which would yield measurable units that he could quantify and convert into equations: his equations, he explains, "*were obtained as the interpretation of the mechanism which I have described in Chapter IV.*"[6] Chapter 4 of *Mathematical Investigations* contains six two-dimensional diagrams that help the reader visualize the mechanism that Fisher was interpreting.[7] The diagrams offer various views of a fluid-filled rectilinear container. Inside it float a number of smaller containers that Fisher calls "cisterns." These cisterns vary in size and shape and are attached to each other by horizontal and vertical rods, which extend from the tops of the cisterns to the sides of the rectilinear container. Two rows of cisterns (on the right and left) represent individual consumers, and two other rows (in the front and the back) represent individual commodities. By means of internal partitions made of wood, each of the cisterns in the front and back rows is divided into two parts: the front part of each cistern represents the physical units in which commodities are valued (pounds, yards, etc.) and the back part represents value in money (dollars). The rods operate in such a way as to "keep the continuous ratio of marginal utilities, the same for all individuals and equal to the ratio of prices."[8] Using a system of valves, tubes, stoppers, and pumps to control the inflow and outgo of water, the operator could use the machine to demonstrate a number of relationships among quantities, between kinds of units, and between quantities and units, including price. These relationships include (but are not limited to) the following: "*the quantities of each commodity consumed by each individual during the year*"; "*the given total quantities of each commodity consumed by the whole market*"; "*the marginal utility of each commodity to each individual*"; "*the prices of commodities in terms of each other*"; and "*the marginal utility of money to each individual.*"[9]

As Fisher described it, the machine his diagrams sketch "is the physical analogue of the ideal economic market." By making all the "elements which contribute to the determination of prices . . . open to the scrutiny of the eye," these diagrams, as well as the literal machine he constructed from them two years later, constituted an "instrument" that Fisher thought could actively enhance investigation. In the operations the mechanism facilitated, Fisher argued, "we are thus enabled not only to obtain a clear and analytical *picture* of the interdependence of the many elements in the causation of prices, but also to employ the mechanism as an instrument of investigation and by it, study some complicated variations which could scarcely successfully be followed without its aid."[10] Without the machine to illustrate the factors that influence prices, in other words, it would have been more difficult not simply to picture their influence upon each other but also to investigate the more "complicated variations" of their interrelations.

The drawings, descriptions, and three-dimensional machine that Fisher presented as "analogues" for an economy seeking equilibrium seem to constitute relatively simple mimetic representations. That is, in helping the student visualize the economy as analogous to a water-filled tub containing a system of linked and differently weighted floating cisterns, these representations seem simply to point to and quantify things that exist in the material world. While some of the things Fisher mentions could be quantified, however, this was not true of all of the items he lists. Thus one could measure and presumably record "the quantities of each commodity consumed by each individual during the year" and, conceivably, "the given total quantities of each commodity consumed by the whole market"; but "the marginal utility of money to each individual" had to be inferred, based on the theoretical assumptions about what makes one "point" or "degree" of utility marginal and why this matters. Then, too, the economy does not exist in the same physical form that Fisher's machine did, cisterns floating in water constitute a poor representation of desire ("utility"), and even the equilibrium that was thought to establish prices was a theoretical assumption about an idealized state that economic processes never actually, or permanently, achieve. The mechanism to which Fisher applied mathematical interpretations, then, existed at one remove, at least, from economic transactions conducted by actual economic actors; and the mathematical equations he used to convey the more "complicated variations" of all of the interrelated factors did not simply refer to or add up actual sales, purchases, prices, or quantities of money in circulation (much less the desire that motivated these transactions) but, instead, created an abstract—but useable—analog for these factors.

Even if this analog could help the reader visualize the interrelationships that connected economic elements, then, neither the drawings nor the machine solved all of the challenges involved in Fisher's ambition to depict the entire economy. His analog machine *was* able to solve two, related problems. By inventing a unit by which the economist could "measure" the movements that stood in for quantities of desire, which Fisher called the "util," he was able to assign numerical values to what other economists typically depicted only with rising or falling lines on graphs.[11] By using the util to quantify the cubic inches by which the movement of the cisterns displaced the water level, he was also able to render the units in which commodities were measured (gallons, tons, yards, etc.) commensurate with the units in which prices were measured (dollars) and thus to calculate the relationship between the movements of the cisterns and the changes in price. Even if Fisher's util solved the problem presented by the incommensurability of the units used in real-world economic activities, however, and even if his mechanism enabled the student to imagine that measurable changes in the water level of the tub were analogous to changes in prices *and* in "levels" of supply and demand, neither the util nor the mechanism could really quantify desire or measure "marginal utility." Nor could Fisher's machine address even more fundamental problems. It could not take account of all of the variables that factored into real-world transactions, and it could not use the data that was available about these transactions. In order to solve these problems, Fisher took another step in the five-year journey we are describing. He dropped the method he used in *Mathematical Investigations*, in which mathematical calculations were used primarily to illustrate an a priori theory, and he turned to a method by which actual buyers and sellers had long calculated value in the real world. Using some of the insights his drawings and mechanism had given him, but relying on this much older method to represent economic processes, Fisher discovered that he could address questions about value in a way newly responsive to what actually occurred in the marketplace.

Definitions, Tables, and Accounting Measurement: From Defining "Capital" to Scrubbing Data

In the next important work from this period, Fisher took what may seem like an intellectual detour: he surveyed the existing economic literature in an attempt to figure out why the concept of "capital," which had long been central to economic theory, remained "obscure."[12] Tracing this obscurity to Adam Smith's discussion of "stock" in *Wealth of Nations*, Fisher insisted that the meaning of "capital" should be generalized beyond the restricted eighteenth-century notion of "stock" and turn instead on what Fisher called the "time element." In a real-world context, this time element was both captured and affected by changes in the prevailing interest rate. Viewed in this real-world context, Fisher explained, capital should be understood not only as the form it took at a single moment in time, where it appeared as a "stock" or "fund," but also in terms of the expected "flow" of income this fund would yield, in the form of the incremental payments of interest it would return over time. The concept of "capital" should thus include the capitalization of the income, or services, a present sum would generate as time passed and in relation to the rate of interest.[13] While Fisher's insight about capital was formulated in a preliminary way in this essay, his foray into the most basic terminology of his discipline constituted two conceptual breakthroughs. First, it signaled his recognition that, in order to become a "science," economics had to adopt a set of foundational terms with stable definitions that could serve as the building blocks for theory. And second, the answer he gave to the question posed in the article's title—"What Is Capital?"—demonstrated that he had turned his attention away from what the util had tried but failed to measure—desire—to a new, equally subjective, but potentially more quantifiable object: expectation. If capital was composed not simply of a static "fund" in the present but also of the "flow" of income or services the future would yield in the form of interest, then what the economist needed to measure was not only desire but also expectation.

In *Appreciation and Interest*, the book Fisher published in 1896, he put his new insight to work. Instead of trying to represent the entire economy with an analog machine, as he had done in his dissertation, Fisher engaged what initially looks like a more local problem: he tried to devise a method for representing the relationship between the prices of commodities and monetary value. In the 1890s, understanding this relationship seemed particularly important because commodity prices had been falling dramatically for the entire decade and the value of gold kept going up. The question this situation raised—was it fair to demand that debts incurred at one monetary value be repaid in a currency whose value had changed?—had become so divisive by 1896 that politicians like William Jennings Bryan were demanding that the United States abandon the gold standard and embrace bimetallism instead. Even though many economists had tried to explain why commodity prices were falling while gold continued to appreciate, no one had noticed what Fisher considered central: the "influence of monetary appreciation and depreciation on the rate of interest."[14] Fisher's contribution to the bimetallism debate was based on the insight that even though the nominal principal repaid by the debtor could either be inflated or depreciated based on changes in the value of the currency in relation to gold, if the change in the value of money had been anticipated in the interest rate agreed upon by the lender and borrower, no injustice would result. Precisely because interest is paid in a series of interim payments between the time the debt is contracted and the time it is fully repaid, the total amount paid over time could compensate for the change that appreciation or depreciation made in the value of money. As Fisher explained, "A farmer who contracts a mortgage in gold is, *if the interest is properly adjusted*, no worse and no better off than if his contract were in a 'wheat' standard or a 'multiple' standard."[15]

To explain how market participants could "properly" adjust the interest rate so that the total payment of a debt was fair, Fisher introduced two kinds of calculations, which, he argued, actual borrowers and lenders routinely performed. The first, as this reference to "standards" implies, involves an algorithm that allows market participants to relate two sets of values to each other and to move back and forth between the two standards of valuation on a compound basis. Thus, if a debt was contracted in one standard—say, dollars—the farmer had to be able to figure out the value of this debt in wheat, the manufacturer had to convert it to widgets, and the merchant needed to understand it in terms of whatever commodities he sold. Fisher also assumed that every form of capital commanded its own interest rate, which was expressed in its own standard. If someone borrowed wheat, the interest payments for the debt *could* be denominated in bushels—but they did not have to be so denominated because, in the United States, it was (and still is) conventional to convert all standards to the dollar standard. Fisher argued that this is acceptable as long as participants understand how standards behave relative to one another. Fisher's point was that the conventional system of denominating interest rates on debt in the money standard (dollars) could be "just," even if the value of each standard fluctuates, *if* borrowers and lenders clearly understood the relationship between the interest accruing on one standard (the dollar) and that accruing on another (say, wheat). A "simple formula," which allowed market participants to compare two rates of change on a compound basis, enabled borrowers and lenders to calculate such relations between standards.[16]

The second calculation essential to understanding the relationship between interest rates and monetary appreciation (or depreciation) involved the "present value" of the future sum that would be returned on money put out at interest. As Fisher explained, "The ordinary definition of the 'present value' of a given sum due at a future date is 'that sum which put at interest to-day will "amount" to the given sum at that future date.'"[17] Present-value calculations, which had been used routinely in insurance, annuity, and fixed-income (bond) transactions since the seventeenth century, allow market participants who foresee changes in the value of money to make rational business and investment decisions, based on the effects these changes will have on costs and income. Instead of requiring complex, on-the-spot calculations, present value is based on elaborate tables that show the geometrical growth over time that is produced by the compounding of interest. Such tables, Fisher points out, which were readily available, "are constructed on this principle [compounding] for the practical use of insurance companies in calculating their premiums, and for brokers in determining the comparative methods of various bond instruments."[18] Fisher's radical insight about present-value calculations was that this technology also permitted borrowers and lenders to embed their expectations about future changes in the value of money in the interest rate at which they agreed to lend or borrow.

Using these two calculations, Fisher argued, the farmer routinely factored his expectations about future changes in the value of money in relation to his expectations about the yield of his wheat crop into the interest rate he was willing to pay. To the objection that ordinary farmers and merchants could not understand appreciation, make such calculations, or anticipate future changes in the value of various standards, Fisher sharply retorted that market participants only needed to follow the trends of relative price changes, which successful farmers and merchants already did as an ordinary part of making business decisions. As the borrower responded to present changes in the value of money, moreover, he actually affected its future cost, through the market mechanism of supply and demand. The ordinary person's "effort is not to predict the index numbers of Sauerback or Conrad, but so to foresee his own economic future as to make reasonably correct decisions, and in particular to know what he is about when contracting a loan. If gold appreciates in such a way or in such a sense that he expects a shrinking margin of profit, he will be cautious about borrowing unless interest falls; and this very unwillingness to borrow, lessening the demand in the 'money market,' will bring interest down."[19]

Fisher's contributions to the bimetallism debate, then, consisted of two basic insights. First, because capital must be understood not simply as the "fund" available in the present but also as the "flow" of income from the future, in the form of periodically paid interest, interest rates could adjust for the changes in standards caused by appreciation and depreciation. Second, because market participants routinely act on their expectations about the future and their understanding of the principal of compounding when they agree on interest rates, these expectations constitute a critical component of economic processes. Indeed, because the behavior of market participants, which is based on these expectations, influences what happens in the market, such expectations are even more critical for the economist to understand than is the desire that he could only assume drives market demand.

In the first pages of *Appreciation and Interest*, Fisher spelled out the goals he wanted to accomplish in this work: "to develop the theory in a quantitative form, to bring it to a statistical test, and to apply it to current problems, and to the theory of interest."[20] If part I accomplishes the first of these goals and offers a down payment on the third, it was his engagement with the second that has made this book so consequential. For, the desire to bring his theory "to a statistical test" both articulated an ambition Fisher had not expressed in *Mathematical Investigations* and required a mode of representation that the analog machine did not supply. Whereas the machine could help the student imagine equilibrium by means of an analogy between cisterns floating in a tub of water and prices and desire, the calculating technologies Fisher invoked in *Appreciation and Interest* could use actual prices, which recorded real market transactions, to check whether or not his hypothesis that expectations were embedded in interest rates was born out in fact.

That present-value calculations were routinely performed by real-life market participants, even if they only dimly intuited the meaning of "appreciation," constituted a critical component of Fisher's argument. Paradoxically, this means that the mode of representation he sets out here was simultaneously more referential or literal than its counterpart in the analog machine and less so. It was more literal because the present-value calculations the tables made possible occurred every day *and*, as we are about to see, because the principles implicit in these calculations were also embedded in the prices of commodities recorded in contemporary newspapers and thus available to Fisher as the data against which he wanted to test his theory. But this mode of representation was also *less* referential or literal than the analogy embodied in the machine because the algorithm that informed the compound interest tables assumed a mathematical relationship between the numbers the table contained that was based on an old assumption that had become conventional. This mathematical relationship, which showed a geometrical as opposed to a linear progression, was based on the principal of compounding and the assumption that underwrote it: that the value of money was related to time. Whereas the representation of the economy created by Fisher's analog machine existed at one remove from actual economic transactions, then, the representation of financial relationships that Fisher created in *Appreciation and Interest* was simultaneously more empirical *and* more abstract. To explain this complex situation, we need to turn to part II of Fisher's text.

Part II, which is simply entitled "Facts," opens with a statement that signals how far Fisher had traveled since writing *Mathematical Investigations*: "No study of the relation between appreciation and interest would be complete without verification by facts. In imaginary illustrations, such as those used in part I, it is easy to make calculations agree to the last decimal place; but the figures in which we are really interested must come from actual market quotations." Most obviously, Fisher's desire for "verification by facts" seems to repudiate the deductive method of his earlier work. Then, too, one of the theoretical assumptions that does inform *Appreciation and Interest*—the assumption that "business foresight exists"—is an assumption different in kind from the theoretical claim about equilibrium that informs *Mathematical Investigations*. One can simply look at "modern business" to see that buyers and sellers eagerly watch "every chance for gain," and one can find the information market participants use to make predictions about the future in the "multitudes of trade journals and investors' reviews" that periodically supply reams of useful data.[21] By contrast, equilibrium exists only in theory; the util is a purely theoretical tool; and, to represent equilibrium, economists must assume the operation of a desire they can never actually measure.

Even though part II makes it clear that Fisher wanted to test his hypothesis that "an expected change in the value of money has an effect on the rate of interest"[22] against actual data, the data he cites in *Appreciation and Interest* are by no means straightforward or simple. Most obviously, these data, which had been collected for a variety of purposes and which existed in a variety of forms, did not always allow Fisher to perform the calculations he so painstakingly demonstrates in part I of his text. To glimpse the form in which the data actually existed in the 1890s, the reader must look beneath the body of the text to the extensive footnotes that appear on almost every page of part II. For, the neatly organized tables that Fisher includes—of rates of interest expressed in various standards (gold, currency, and coin); rates of interest realized on India bonds; rates of interest in relation to prices in New York, London, Berlin, Paris, Calcutta, Tokyo, and Shanghai; rates of interest in various cities in relation to rising and falling prices and wages; and average bank rates before and after the breakdown of bimetallism—constitute the end products of a set of elaborate procedures that Fisher only acknowledges in the footnotes. These procedures, which Wall Street analysts now call "data scrubbing," were necessary to render the information included in newspaper reports, official documents, and the other sources Fisher consulted commensurate. This process of commensuration, which could also be called "cleansing" or "amending," involved removing incorrect or inconvenient elements from the available data, supplying missing information, and for-matting it so that it fit with other data. Fisher does not mention the labor involved in scrubbing data in the body of his text, but the elaborate footnotes make it clear how "laborious" this work could be. Here, for example, is Fisher's explanation of how he generated an interest rate table for the period 1875 to 1895 that compares India bonds in gold and silver to the estimated and actual percentages of appreciation of gold in silver:

> The methods by which the first column [showing rates of interest in silver] is computed are the same as those explained in the preceding chapter, account being taken of the fact that the price quotations for rupee paper are not "flat," so that no corrections for accrued interest need be applied. For computing the second column [interest rates in gold] a more laborious method was necessary, due to the fact that the quotations are not continuous of the same bond. The earlier ones are for a 4% bond and the later for a 3% bond. The buyer of a 4% bond is regarded as converting it into the 3% at the current price in 1888, the date of maturity of the earlier bond. As no bond tables apply to such conversions, tables of present values were used and that rate was found by trial (and interpolation) which would make the present value of all benefits equal to the purchase price.[23]

Fisher's reference to the absence of bond tables—and to his substitution of present-value calculations for those tables—makes it clear that scrubbing was not the only process that rendered the available numbers commensurate to each other. Beyond reflecting the laborious labor necessary to make the numbers speak to each other, the figures that appear in Fisher's tables also constitute outcomes of the algorithmic operations—and thus the historical, theoretical, and conventional assumptions—represented in present-value tables and in other statistical and mathematical methods. Fisher takes it for granted that the numbers in his tables embody the outcomes of these operations, both because the only numbers that are useful for his tables must take this form and because this is the form in which such figures appear in the price lists, government reports, and investors' guides available to him. Fisher acknowledges this when he casually remarks that he generated the numbers that appear in various tables "by the aid of the usual brokers' bond tables," that he used the statistical method called averaging to make other sets of data work, and that he relied on index numbers to generate yet another table.[24] All of these methods—algorithmic, statistical, and mathematical— were commonly used in Fisher's day, of course, and there is nothing remarkable in his use of them. Our point is simply that even an economist who wanted to do empirical work, who wanted to take the data available in newspapers and trade journals as his sources, always carried over the assumptions implicit in the ways these data were collected and presented in the first place, even as his use of them effaced their historical and conventional nature.

Conclusions

Now we can see why Fisher abandoned the mechanical-analogical method featured in his earlier work and why this change in representational mode has proved to be so consequential. The util, which is the unit of measurement he created to facilitate commensuration in his analog-machine method, could *only* represent economic processes analogically—only, that is, by producing a simulation that reproduced the theoretical assumptions formulated as equilibrium theory. This method could not quantify desire, in other words. What the analog actually depicted was simply the theoretical assumption with which Fisher began: in an idealized economy, the equilibrium between desire/demand and supply establishes prices. The mode of representation Fisher adopted in *Appreciation and Interest*, by contrast, could refer to real-world transactions because the prices he quoted were those actually paid by market participants and because the algorithm he used to express his theory was the *same* algorithm used in those transactions. Indeed, this mode of representation was simultaneously mathematical, empirical, conventional, *and* theoretical: In marrying theory to empirical data in a quantitative form, Fisher created a framework that could fit the theory to the data—but only if various kinds of data were made commensurate with each other and only if the data embedded the assumptions that made them useable in the first place.

The labor involved in rendering the available data commensurate, which Fisher describes in the footnotes to *Appreciation and Interest*, might seem to undermine even the tentative conclusions Fisher presents in this work. But it is important to acknowledge how novel Fisher's ambition was in the late nineteenth century. Almost all of his peers in the profession, even those who celebrated empiricism and fact-based work, tended simply to gesture toward what data *might* prove, *if* it were to ever to be available.[25] As William J. Baumol has tartly remarked, the method of almost all late nineteenth-century American economists relied primarily on "categorical pronouncements based on no more than personal conviction."[26] Fisher's data might seem to us to be both uneven and unduly laundered, but his determination to use what data existed and his painstaking descriptions of how he made it useable foretold the directions economics would eventually take.

Even though data entered Fisher's work somewhat belatedly, moreover—at least relative to his drawings and machines—what his treatment of data reveals is that the economist's data always exists at several degrees of remove from the world of actual market transactions. The numbers that prove useful for the economist's calculations, in other words, already embed a set of conventional assumptions, which are simultaneously reinforced and effaced in technologies like present-value calculations and compound interest tables. When the economist scrubs such data to make it more amenable to the calculations he wants to make, he repeats a process of elaboration and obfuscation that is already implicit in them. Fisher's two crucial contributions to modern economics turn on his *use* of such assumptions—not on anything he did to call attention to them. First, Fisher realized that what the economist needed to quantify was not the force of desire but expectations about the future; in doing so, he helped make a future-oriented understanding of value central to economic theory. And second, because he recognized the usefulness of the present-value calculations by which real-world actors routinely represented these expectations, he reinforced a process long underway: the naturalization of an old set of assumptions about money and value. Irving Fisher thus helped make financial economics both a predictive science and one that could justifiably claim to draw its inspiration from the past and to base its predictions on empirical data. As a precursor of modern *financial modeling*, *Appreciation and Interest* reminds us that even Fisher's peculiar treatment of data made—and, to most modern economists, still makes—sense.

Notes

1. Irving Fisher, "Mathematical Investigations in the Theory of Value and Price," dissertation, Yale University mathematics department, 1891. Reprinted in 1926 and 1965. All references to this work, abbreviated in these notes as *MI*, refer to the 1965 reprint (New York: Augustus M. Kelley). This volume also contains *Appreciation and Interest*, which we discuss subsequently. The pagination of this latter text is self-contained; we cite page numbers from this version of *Appreciation and Interest*, as *AI*, in our text.

2. The only scholar to recognize the significance of *Appreciation and Interest* in terms that resemble ours is Robert W. Dimand. See Dimand, "Irving Fisher and the Fisher Relation: Setting the Record Straight," *Canadian Journal of Economics* 32, no. 3 (May 1999): 744–750.

3. "Desire" is Fisher's term: "No one ever denied that economic acts have the invariable antecedent, desire" (Fisher, *MI*, 11). Equilibrium theory also contains the idea of marginal utility, which states that an individual's desire for a commodity or service (its "utility") decreases as the

individual acquires more of that commodity or service. Since price is a function of utility, the price decreases as the quantity reaches its marginal level. Fisher's invented measurement unit, the util, was derived from "utility."

4. Fisher, *MI*, 3.

5. Ibid.

6. Ibid., 4.

7. The 1926 edition of Fisher's PhD thesis includes not only Fisher's original diagrams but also photographs of both physical constructions of Fisher's model. The 1893 model is the smaller of the two; its container is a box set on what appear to be sawhorse supports. The model built in 1925, by which time the first model had worn out from repeated classroom use, seems to fill an entire room. Fisher never discussed the complexities introduced when he converted his two-dimensional diagrams into a three-dimensional machine. Nor does he belabor the precision (or lack thereof) of the fit between the diagram and the economic principles it helped him visualize. The former is the subject of Mary S. Morgan's and Marcel Boumans's treatment of another hydraulic machine, built by yet another economist (Bill Phillips): "Secrets Hidden by Two-Dimensionality: The Economy as a Hydraulic Machine," in *Models: The Third Dimension of Science*, ed. Soraya de Chadarevian and Nick Hopwood (Stanford, CA: Stanford University Press, 2004), 369–401. The latter is the subject of Mary S. Morgan's "The Technology of Analogical Models: Irving Fisher's Monetary Worlds," *Philosophy of Science* 64, Supplement Proceedings of the 1996 Biennial Meetings of the Philosophy of Science Association, Part II: Symposium Papers (December 1997): S304–S314.

8. Fisher, *MI*, 40.

9. Ibid., 42–43.

10. Ibid., 44.

11. These lines are the supply and demand curves, whose intersection enables economists to represent how markets determine price, according to the theory of equilibrium.

12. "What Is Capital?" *The Economic Journal* 6, no. 24 (December 1896): 509–534.

13. Ibid., 520, 525, 526. Fisher returned to this subject repeatedly, most notably in *The Nature of Capital and Income* (New York: Macmillan, 1906). The assumption that money *should* earn interest when it is lent out is based on the principle modern economists call the time value of money. Historically, this principle derives from very old Christian notions about the relationship between time and God; theorists who wanted to circumvent the Church's ban on usury elaborated on it in the sixteenth century. One of the economic theorists whose work on this subject Fisher specifically engaged in *Appreciation and Interest* was Eugene von Bohm-Bawerk, the Austrian finance minister. Fisher's most important departures from Bohm-Bawerk's *Capital and Interest*

(1890) followed from Fisher's decision *not* to reproduce the historical account of interest that the Austrian patiently constructed. Effacing the history of the assumption that money lent out over a period of time should accrue interest has played a critical role in the naturalization of modern assumptions about capital.

14. This is part of the title of Fisher's text in its initial publication form, which appeared in *Publications of the American Economic Association* 11, no. 4 (August 1996): 331–442.

15. Fisher, *AI*, 16; emphasis in original.

16. Ibid., 9.

17. Ibid., 19.

18. Ibid. Fisher does not bother to prove the "general theorems" that present-value calculations present because, as he explains, "their proof is accessible in most treatises on interest, annuities, insurance, etc." (*AI*, 20, n. 1). So numerous were the texts that proved and displayed these "elaborate tables" that Fisher merely gestures, here and elsewhere, to the sources he clearly assumed his reader would know: "See, e.g., the 'Encyclopedia Britannica,' 'Annuities'" (*AI*, 20, n. 1). Fisher also refers to "Horner's Method" (*AI*, 27), which was a technique of synthetic division used to evaluate polynomials. Horner's Table (or Tableau) provided a shortcut so that one would not have to make the laborious calculations necessary to produce these results. Fisher was intimately familiar with such tables and their usefulness; in 1894 he and a Yale colleague, Andrew W. Phillips, published a five-figure table of logarithms to accompany the geometry textbook that Fisher and Phillips published in 1896.

19. Ibid., 36.

20. Ibid., 5.

21. Ibid., 35, 37.

22. Ibid., 38.

23. Ibid., 50, n. 6.

24. Ibid., 47, 59.

25. Foremost among late nineteenth-century economists who celebrated empiricism but nevertheless took a polemical stance, regardless of the data, was J. L. Laughlin at the University of Chicago.

26. "On Method in U.S. Economics a Century Earlier," *American Economic Review* 75, no. 6 (December 1985): 1–12, at 3. Baumol's examples include Richard Ely and Arthur Twining Hadley (president of Yale and Fisher's colleague). Baumol argues that, as a group, economists did not embrace empiricism, mathematics, or statistics until World War I.

4 Where Is That Moon, Anyway? The Problem of Interpreting Historical Solar Eclipse Observations

Matthew Stanley

Auguste Comte, the founder of positivism, placed astronomy at the top of his hierarchy of sciences.[1] This was in part due to the simplicity of data in positional astronomy—usually just a pair of angles—leaving no room for contamination by bias or subjectivity.[2] Consider a total solar eclipse, the stunningly precise predictions of which are an amazing testament to the power of astronomy. The data characterizing such an eclipse has two parts: a time and a place, such as "May 25 585 BC, 10:35 AM, at Babylon." Scientific observations do not come much more raw than this.

But the raw data is not so raw. Acquiring this simple information for eclipses in the past might seem straightforward, but is actually the result of complicated, messy processes that are far from standardized. Even when an eclipse's time and place are explicitly recorded in, say, a Babylonian tablet, they can only become data after astronomers grapple with a number of literary, historical, and psychological factors. What might seem to be the most basic observation imaginable—when was the sky dark?—actually became dependent on highly contested processes relying on methods far from the expertise of astronomers.

The Problem

The dating of ancient solar eclipses is certainly of historical interest, as they can help establish chronological benchmarks. But the eclipses are of great importance for purely technical astronomy as well. Astronomy prides itself on its ability to predict the motions of celestial bodies accurately, but the closest of these bodies—the moon—is actually one of the most recalcitrant. Its complicated motions continually cause headaches and, even worse, seem to have changed over time. That is, the equations that describe its motion work very well for predicting eclipses in the present, but do not seem to work as well for eclipses in the past. When astronomers try to place the moon at a date in the past when there was known to be an eclipse, their equations often say there was no such eclipse, or place the eclipse at a different location from where it was observed. The easiest solution to this problem is to add a small term to the equations of the moon's motion that will place the eclipse in the right time and place. The change in motion is called the "secular acceleration."

The existence of this acceleration was first suspected by Edmund Halley in 1692, and was confirmed by astronomers in the eighteenth century.[3] An accurate value for this acceleration is quite important, because without it, it becomes very difficult to predict the moon's motion more than a century or two away from the present. Unfortunately calculating this value is rather challenging, as one needs an accurate location for the moon in the distant past. Reports of historical eclipses provide exactly this data—when and where the eclipse was seen. Using historical records proved to be problematic, however, and there was significant disagreement about how best to go about it.

The Nineteenth Century

Astronomers had begun working with ancient eclipses to determine the secular acceleration in the late seventeenth and eighteenth centuries. In the nineteenth century they became more reflective of the difficulties involved in using historical sources for precise calculations. In the middle of the century, the Astronomer Royal George Airy argued that Laplace's powerful lunar theory and Hansen's new tables made obsolete all previous work, and demanded a recalculation of the secular acceleration. In this process he raised a simple but profound problem. In a solar eclipse the moon can cover varying amounts of the solar disk (characterized as the magnitude of the eclipse), and different magnitudes indicate different locations for the moon. Building on earlier work by the famous eclipse hunter Francis Baily, Airy argued that one well-known eclipse must have been total, thus giving even more precise data than previously available. This eclipse, known as the eclipse of Thales, was reported by the historian Herodotus as having stilled a great battle in Asia Minor, which Airy drew out as evidence for totality: "I have myself seen two total eclipses (those of 1842 and of 1851), being on both occasions in the open country; and I can fully testify to the sudden and awful effect of a total eclipse. I have seen many large partial eclipses, and one annular eclipse concealed by clouds; and I believe that a body of men, intent on military movements, would scarcely have remarked on these occasions anything unusual."[4]

Airy argued that his personal experience of eclipses was necessary for interpreting the eclipse of Thales as a total one. In particular, it was only the "sudden and awful" experience of seeing a total eclipse that could have disrupted a pitched battle. Similarly, a reference in Xenophon to the surrender of the besieged city of Larissa when a cloud covered the sun was interpreted as a total solar eclipse, that being the only way to explain the terror of the well-fortified citizens.[5] Thus the establishment of the fact of an eclipse's magnitude was dependent on claims about the psychology of seeing such a phenomenon.

Airy was convinced that the texts containing these reports were highly reliable. An amateur classicist, he took "great enjoyment" in studying Greek texts, and reported that as a student he spent as much time reading the classics as he did mathematics.[6] He read Herodotus and the other Greek authors as realistic history, a nearly universal practice in the Victorian period.[7] For example, he spent significant time analyzing the eclipse seen by Agathocles as the tyrant fled Syracuse. The eclipse seemed promising for astronomical use except that the text did not tell whether Agathocles sailed north or south from Syracuse, thus obscuring the critical fact of where the eclipse was seen. Airy thought he could solve this problem through deep historical investigation, including classical naval strategy, the logistics of armies, and the proper identification of certain North African quarries. He was convinced that his classical sleuthing had revealed Agathocles' route, and thus that he had achieved "perfect certainty" regarding the data of this eclipse.[8] Airy's confidence in the accuracy of these ancient texts was even more remarkable given his well-known obsession with precision and highly disciplined observing regimes.[9]

Not all astronomers were willing to credit these moves as producing reliable data, however. In the American astronomer Simon Newcomb's lunar theory "the ancient total eclipses of the sun, which have been so much discussed during the present century, are here thrown aside."[10] Newcomb's journey to professional astronomy was quite different from Airy's classics-steeped gentlemanly education. Virtually self-taught, as a youth he literally walked out of the woods of Nova Scotia and managed to impress the astronomical community with his extraordinary skill in calculation. Known for his manic energy and "massive head," he wrote extensively on the need to bring scientific methods to bear on all aspects of society.[11] While he was a close friend of Airy's, he felt strongly that science was far above all other forms of human activity, and could not be contaminated by insecure historicism.

Newcomb began his criticism of Airy's use of Greek sources slowly. Perhaps, he said, Airy was not justified in assuming that all eclipses in the ancient chronicles were total.

Certainly he could not "conceive that the historic evidence bearing on the subject places the phenomena of totality so far beyond doubt that a discussion of other data is unnecessary."[12] He argued that even if the inference of totality from reactions recorded in the historical texts was valid, the texts themselves might not be reliable. The statements were "so vague that they may be referred to other less rare phenomena. It must never be forgotten that we are dealing with an age when accurate observations and descriptions of natural phenomena were unknown, and when mankind was subject to be imposed upon by imaginary wonders and prodigies."[13] In a time of mythology and epic tales, we should not expect writers to adhere to modern standards of evidence. Perhaps, Newcomb said, a reference to stars being seen in the daytime could indicate a total eclipse. But even then, the visibility of Venus could make this difficult.

Regarding the eclipse of Thales, the tale as written would "hardly even suggest an eclipse of the sun, or anything more extraordinary than the regular advent of night, except for the single word ἐξαπίνης (suddenly)." Perhaps during the fury of battle, the combatants lost track of time and were surprised to realize that night had fallen only when it interfered with the fight. Or even more simply, a dark cloud. The story did not sound as remarkable to Newcomb as Airy and others suggested: "If it be urged that the making of peace indicated something extraordinary or impressive, we may rejoin that there is nothing in the account to indicate it." Further, a total eclipse would probably have been too quick to alter the course of a battle: "if the phenomenon was really that of a total eclipse, the night must have turned back to day again almost before the fighting could stop, a fact which the historian does not mention."[14]

Similarly, Airy had claimed that it "cannot be doubted" that the darkness at Larissa (detailed in a passage translated by him) was actually an eclipse. Newcomb laconically said that he was "unable to share the confidence of the Astronomer Royal." The writers of the time were simply not reliable: "The narratives of these times contain many accounts of wonderful occurrences, in which we know that a liberal allowance is to be made for the flight of the imagination; and it is not entirely logical to accept unhesitatingly all those statements which we can reconcile with our knowledge, while we reject all others."[15] The problem, Newcomb said, was that proper observation of the natural world could only be the result of rigorous scientific training, thus accounting for "the rude and doubtful character of nearly all the ancient data."[16] So "the uncritical character of Herodotus" was not particularly his fault.[17] An eclipse reported by Thucydides was more reliable, given the report that stars were also seen. This was likely an eclipse, but again "not sufficient to justify the introduction of an equation founded on it."[18]

Similarly, recently discovered Chinese and Assyrian eclipse observations were of no value. Their use would have relied on "the judgment of the investigator. It would hardly have been possible to have formed such a judgment without the suspicion of its being influenced by his wishes or prejudices."[19] Such subjectivity was exactly what Newcomb thought made Airy's classicist analysis unreliable, and it had no place in astronomy.

Newcomb's skepticism was challenged by P. H. Cowell, a chief assistant at the Royal Observatory Greenwich. Cowell insisted that a careful reader could tell when a historical record was that of a total eclipse, because "there are words in the record implying at least a near approach to totality . . . fire in the midst of heaven . . . day turned into night . . . stars seen . . . light nearly extinguished." He also claimed that individual historical writers needed to be given credit for their reputations for personal reliability: "Professor Newcomb does not consider . . . the character of the historian for accuracy, as inferred from his other writings."[20] Cowell even made an argument for the totality of historical eclipses from the paucity of records. Astronomers know that there are few total eclipses, but many partial eclipses; there are few records of eclipses at all; therefore, the records we have must be of total eclipses, or there would be many more.[21] He then went on to use these total eclipses as reliable data.

Newcomb had also disputed that we knew where the eclipse was seen, which was vital for turning a historical record into astronomical data. For many cases, Airy had simply assumed that the eclipse was located where the writer lived, which Newcomb found fairly bizarre.[22] He further refused to go along with Cowell in trying to guess where a writer such as Thucydides happened to be traveling when seeing an eclipse. Cowell maintained it was "reasonable to impose the condition that Thucydides was not at any rate further off than the Bosphorus . . . I however see no reason to doubt that Thucydides was in Athens . . . We know he was in Athens the following summer, when he caught the plague. . . . But he might, it is said, have gone to look after his estates in Thrace."[23]

The Twentieth Century

The task of reevaluating the data apparently hidden within historical eclipse records seems to have been taken on about once a generation, though not always (as we have seen) with unanimity among astronomers. This project was undertaken again in the 1970s by the American astronomer Robert Russell Newton from Johns Hopkins University. Newton, an expert in satellite navigation systems, began work on the secular acceleration and hoped to use the data provided by Ptolemy in his famous *Almagest*. After uncovering internal inconsistencies in that text (which had been well known for centuries), Newton devoted years to revealing the "crime" of Ptolemy, who was accused of deliberately falsifying data: "It is clear that Ptolemy knew what he was doing."[24]

This became part of a larger crusade against the way contemporary astronomers used ancient eclipse records. He criticized previous investigators as having decided in advance what the secular acceleration must be and choosing only eclipses that matched their presumption. He warned that many ancient references are "ambiguous, and many were almost surely not based on valid observations."[25] Newton argued that he would be able to discern which were reliable by studying the ancient texts "from the standpoint of the texts themselves, their historical settings, and other relevant considerations."[26]

His strategy was to assign each eclipse a reliability number that would be used to weight its data before it was used in any calculations. This number would be assigned on the basis of "textual criticism, a task for which I am not well prepared. However, there is no standard interpretation that can be used for many of the eclipses because there is frequent disagreement among the authorities who have studied the texts. Therefore I have had to carry out independent textual criticism for this study." This criticism had to include the known habits of the ancient writer, the distance in time between the event and the writing, and the tendency to dramatize past events.[27]

Newton proceeded to wade brutally through the mass of eclipse records, discarding some as retroactive prophecies, some as magical, some as literary. Perhaps astronomers were simply reading an ancient equivalent to *A Connecticut Yankee in King Arthur's Court*, and the eclipse was wholly invented.[28] "Magical" eclipses were placed in texts simply to contribute a sense of awe to events, such as one reported in the Gospel of Luke during the crucifixion. Given that a solar eclipse is physically impossible a week after Passover, Newton dismissed this as just one more example of people's "remarkable tendency to die during eclipses."[29]

He was skeptical that the terms used in Babylonian reports (e.g., "fire in the midst of heaven") actually refer to eclipses, or that Xenophon witnessed one. He concluded that the darkness was clearly caused by a cloud, and that "Only romance could call this a useable eclipse record, and I shall not calculate it." Livy's reported eclipses were discarded as magical, since about half of them were accompanied by rains of stones.[30]

The plentiful Babylonian records also came under suspicion. There were too many assumptions underlying the conclusion of their accuracy. First, that human beings record information in predictable ways: there "are assumptions about uniformity of human conduct and uniformity of word usage at all times in all contexts that go beyond my experiences of human consistency." Second, that "the cuneiform text in question can be read with no ambiguity in the technical terms involved. We can test this corollary by comparing independent translations of the (transcription of the) cuneiform record."[31] Newton concluded that the vagaries of translation and linguistics did not allow a high reliability to be assigned. Babylonian astronomical texts had another peculiar problem, which was that it was unclear whether statements such as "eclipse on the first day of the third month" were *observations* of eclipses or *predictions*, thus making them wholly useless for this purpose.[32]

Even further, he denied that most of these eclipses were total—even when explicitly described as such. One could only distinguish between a total and annular eclipses if one is "an expert astronomer." It seemed that Ptolemy and other ancient astronomers actually did not know about annular eclipses, so their word could not be trusted.[33]

Much of Newton's work was rebutted by F. Richard Stephenson (a British astronomer with extensive expertise in Chinese texts) and Paul M. Muller (formerly of the Jet Propulsion Laboratory) who nonetheless acknowledged the difficulties:

> Since the subjective impression of an eclipse on the part of a human observer is the data to be used in this investigation, there is no definable and repeatable set of experiments which will *prove* the conclusions we are about to give in this matter. In that sense, this investigation must necessarily be somewhat "unscientific." We, and others before us, have been convinced that at least some of the admittedly imprecise and often emotional records retained through history by eye-witnesses to large eclipses can, nevertheless, constitute highly reliable and usable scientific observations.[34]

The pair relied on the "the subjective impressions which arise from total, and near-total, eclipses." The psychological impact of totality, even compared to 0.99% of totality, was so tremendous that it could be relied on. This also had a useful side effect in that they did not have to worry about the training or knowledge of the observer.[35] They could not rely on modern instrumentation or data collection, only raw human experience. What was needed was to think about those ancient observers as instruments: "to make use of these data as scientific observations requires that we calibrate and understand the observers as human beings, and see the kind of events in the context of what the impact would be on these observers."

The observers, long dead, needed to be calibrated like a thermometer. Astronomers were hardly trained in such a task, and had to draw on various other disciplines for it to work: "This kind of analysis cannot completely satisfy the usual scientific requirement of repeatable and definable experimental testing. The diverse disciplines of history, psychology, physiology, common sense, personal experience, and professional judgment enter in complex ways." Astronomical data could not be produced solely by astronomical methods.

Exactly how to deploy these disciplines to produce reliable data was by no means clear. One of the issues with Chinese records, which otherwise were quite useful, was that the records were compiled in the capital but did not reference the location of the observer. China, even in pre-modern times, was a vast territory, and without a precise location the eclipse reports were useless. Newton criticized Muller and Stephenson for assuming that all the observations were made in the capital itself because the emperor would not have been interested in reports from the provinces. To him, "this reads like a statement about how its writer would act if he became the emperor of China without changing his background or personality."[36]

One point of contention concerned a text by the Roman writer Ennius, as quoted by Cicero. Muller used a mention of the moon standing in front of the sun as indication of an eclipse to establish its date and location. Newton critiqued that use, claiming that the dating was based on an error in translation: "The verb ostitit is singular and it can have only one subject. The subject is clearly the moon (luna), and 'night' (nox) clearly belongs to the clause that follows 'and' (et); the poet has given us this clause only in ellipsis. An ellipsis at this point does not make sense if the missing predicate is independent of what has gone before."[37] The text could only become data through a proper knowledge of Latin grammar.

Conclusion

The goal of all of these struggles was obtaining a number: the secular acceleration, which would then modify the equations of the moon's motion. To get this number, one needed other numbers: the time and place of ancient eclipses. But this data only existed once it had been passed through textual, historical, and psychological filters. Each filter could be used positively or negatively, to either exclude a record from reliability or to detect reliable records.

Textual considerations were based on the documents themselves. The difficulty here was that few of the records consulted actually used the word "eclipse" (and in some cases that concept was not even available to the writers). Astronomers looked for words or phrases that seemed likely to represent a total eclipse, such as "day turned into night" or the sun being "extinguished," to filter records in. Others critiqued this practice by emphasizing the vagueness of many passages, the uncertainty of the kind of text (e.g., prophetic or magical), or even the reader's ability to discern an observation from a prediction.

Historical considerations drew on what was known or could be inferred about the context of the writer and the writing process. Airy and Cowell insisted that ancient writers were reliable as a matter of course, and it simply took educated historical investigation to turn Herodotus into data. In turn, Newcomb insisted that these writers were not trained as astronomers and therefore could not be trusted to provide the raw observations needed. Newton argued that without knowledge of how Imperial Chinese records were gathered and compiled, those records could not be considered reliable. Context could push a record either in or out of the category of reliability.

Psychological considerations stand apart in that they were almost always used as a positive filter, that is, as justification for the reliability of the ancient records. Astronomers considered the psychic impact of the sudden darkness of an eclipse, often drawing on their own experience, to determine whether a writer has actually seen totality. Was it terrifying enough to end a pitched battle? Then it was a real eclipse. A very strong line was drawn between total and near-total eclipses in terms of their psychological impact, thus enabling modern astronomers to infer the mental state of writers thousands of years ago. Here, subjectivity was a *benefit*—the personal, emotional reaction to darkness was what marked a record as real data.

The problem of the secular acceleration is strangely persistent. It is rare for the value of a well-defined scientific quantity to remain in dispute for such a long time—in this case, hundreds of years. Part of the problem is surely the difficulty of accumulating new or better evidence. Cuneiform tablets are only unearthed infrequently, and no amount of clever experiments will uncover a new ancient Chinese text. But more severe than this is the difficulty of coming to agreement about the evidence already in hand. Astronomers as a group are simply not trained in the skills necessary for understanding these sources. Without this common background, it is extremely hard to achieve the consensus that underlies, say, the mass of the electron. Until Sumerian is taught alongside celestial mechanics in graduate programs, it is unlikely that the secular acceleration will become settled knowledge.

These controversies, of course, are rarely made public in the published data. A table of ancient eclipses appears to a casual reader as a simple chart of numbers indicating time and place: May 25 585 BC, 10:35 AM, at Babylon. They have all the crispness and precision one expects from astronomy. But in the history of those numbers lies hidden significant intellectual and interdisciplinary struggle whose outcome was far from unanimous within the astronomical community, and continues to bedevil the calculation of lunar motions today.

Notes

1. A. Comte, *The Essential Writings* (New Brunswick, NJ: Transaction Publishers, 2004).

2. Ibid.

3. For positional astronomy in this period in general, see Bruno Morando, "The Golden Age of Celestial Mechanics," *General History of Astronomy, Volume 2: Planetary Astronomy from the Renaissance to the Rise of Astrophysics, Part B: The Eighteenth and Nineteenth Centuries*, ed. René Taton and Curtis Wilson (Cambridge, UK: Cambridge University Press), 211–239. On the moon's secular acceleration, see J. M. Steele, "Dunthorne, Mayer, and Lalande on the Secular Acceleration of the Moon," *Ptolemy in Perspective, Archimedes* 23 (2010): 203–215; and D. Kushner, "The Controversy Surrounding the Secular Acceleration of the Moon's Mean Motion," *Archive for History of Exact Sciences* 39, no. 4 (1989): 291–316. Laplace's influential treatment of these issues can be found in P. S. Laplace, *Celestial Mechanics* (Bronx, NY: Chelsea Publishing Company, 1966–1969).

4. G. B. Airy, "On the Eclipses of Agathocles, Thales, and Xerxes," *Philosophical Transactions of the Royal Society of London* 143 (January 1, 1853): 179–200.

5. "The Astronomer Royal, on Ancient Eclipses," *Monthly Notices of the Royal Astronomical Society* 17 (1857): 233–235.

6. G. B. Airy, *Autobiography* (Cambridge, UK: University of Cambridge Press, 1896).

7. F. Turner, *Contesting Cultural Authority* (Cambridge, UK: Cambridge University Press, 1993).

8. Airy, "On the Eclipses of Agathocles, Thales, and Xerxes."

9. S. Schaffer, "Astronomers Mark Time: Discipline and the Personal Equation," *Science in Context* 2 (1988): 115–145.

10. S. Newcomb, *Researches on the Motion of the Moon* (Washington, DC: Government Printing Office, 1878).

11. A. E. Moyer, *A Scientist's Voice in American Culture: Simon Newcomb and the Rhetoric of Scientific Method* (Berkeley: University of California Press, 1992), 66–67.

12. Newcomb, *Researches on the Motion of the Moon*.

13. Ibid.

14. Ibid.

15. Ibid.

16. S. Newcomb, *Astronomy for Schools and Colleges* (New York: Henry Holt, 1879).

17. S. Newcomb, *Popular Astronomy, School Edition* (New York: Harper and Brothers, 1896).

18. S. Newcomb, *Researches on the Motion of the Moon* (Washington, DC: Government Printing Office, 1878).

19. Ibid.

20. P. H. Cowell, "Note on the Astronomical Value of Ancient Eclipses, Reply to the Above Note," *Monthly Notices of the Royal Astronomical Society* 66 (1905): 35–36.

21. Ibid.

22. S. Newcomb, *Researches on the Motion of the Moon*.

23. P. H. Cowell, "On Ancient Eclipses," *Monthly Notices of the Royal Astronomical Society* 66 (1906): 523–542. Cowell wrote numerous letters to Newcomb begging the American to be more flexible in matters of interpretation. See Container 20, Simon Newcomb Papers, Manuscript Division, Library of Congress, Washington, DC.

24. R. R. Newton, *Ancient Planetary Observations and the Validity of Ephemeris Time* (Baltimore, MD: Johns Hopkins University Press, 1976) and *The Crime of Claudius Ptolemy* (Baltimore, MD: Johns Hopkins University Press, 1977).

25. R. R. Newton, *Ancient Astronomical Observations and the Accelerations of the Earth and Moon and Medieval Chronicles and the Rotation of the Earth* (Baltimore, MD: Johns Hopkins University Press, 1970).

26. Ibid.

27. Ibid.

28. R. R. Newton, "Two Uses of Ancient Astronomy," *Philosophical Transactions of the Royal Society of London* A 276 (1974): 99–110.

29. R. R. Newton, *The Moon's Acceleration and Its Physical Origins, Volume 1, as Deduced from Solar Eclipses* (Baltimore, MD: Johns Hopkins University Press, 1979).

30. Newton, *Ancient Astronomical Observations*.

31. Ibid.

32. Newton, *Ancient Planetary Observations*.

33. Newton, "Two Uses of Ancient Astronomy."

34. P. M. Muller and F. R. Stephenson, "The Accelerations of the Earth and Moon from Early Astronomical Observations," in *Growth Rhythms and the History of the Earth's Rotation*, ed. G. D. Rosenberg and S. K. Runcorn (New York: Wiley Press, 1975), 459–534.

35. Ibid.

36. Newton, *The Moon's Acceleration*.

37. Ibid.

5 "*facts* and FACTS": Abolitionists' Database Innovations

Ellen Gruber Garvey

It is well known that *American Slavery As It Is: Testimony of a Thousand Witnesses* had a tremendous impact on the U.S. abolition movement when it was published by the American Anti-Slavery Society in 1839. Produced through the collaboration of Angelina Grimké Weld, her husband, Theodore Weld, and her sister, Sarah Grimké, the book offered American abolitionists new ammunition for their spoken and written war against slavery.[1] What is less well known, however, is that *American Slavery As It Is* was the product of a new way of using media, one that is now familiar to us through our computer-based keyword and Lexis/Nexis searches. The book combined personal testimony from those who lived, or who had lived, in the South, some of them former slaveholders, elicited via a form letter—a questionnaire of sorts—with evidence gleaned from a vast archive of newspapers. Here I will focus on that innovative use of newspapers, for in writing *American Slavery As It Is*, the Grimkés and Weld reconceptualized the press to mine it as a database, and modeled ways other abolitionists could use the press and the writings of the South against itself.[2]

Sarah and Angelina Grimké were born into a slaveholding family in South Carolina but rejected that life to become ardent abolitionists, traveling New England as accomplished, convincing speakers, testifying to their direct experience of seeing the effects of slavery on both slaves and owners. They drew on their experience in their writings as well. (Angelina Grimké wrote the only antislavery work by a Southern white woman addressed to other Southern women, *An Appeal to Christian Women of the South*, 1836). When Angelina Grimké married the abolitionist and reformer Theodore Dwight Weld in 1838, both were in frail health. They settled in Fort Lee, New Jersey, with Angelina's sister, and all three retired from public speaking. Abolitionist friends were dismayed at losing such effective orators. The three next took up an extraordinary work, *American Slavery As It Is*, the most widely read antislavery publication until the novel *Uncle Tom's*

Cabin was published serially in 1851. Indeed, Harriet Beecher Stowe reported that she kept *American Slavery As It Is* "in her work basket by day and slept with it under her pillow at night till its facts crystallized into Uncle Tom."[3] *American Slavery As It Is* was priced to sell in bulk for widespread distribution—37 1/2 cents each, and $25/hundred—and sold a hundred thousand copies in its first year.[4] It was an important gesture in the move away from theology or exhortation, and toward reliance on documented, factual information to change the minds of white Northerners about slavery. As Angelina Grimké Weld wrote to her sister Anna R. Frost, "facts, FACTS, have set in motion all that machinery in England" that freed the slaves in the British West Indies and turned England against slave-grown cotton.[5] The English abolitionists had discovered that compiling concrete facts and statistics—such as the high percentage of British sailors who perished on slave ships, gleaned by patiently combing through ships' logs—was far more effective in turning public opinion than appeals to sentiment. Data will out.

American Slavery As It Is compiled testimony from those who had lived in the South and from former slaveholders like Sarah and Angelina Grimké themselves, but it also relied heavily on materials from the Southern press, particularly advertisements for runaway slaves. Such ads had appeared in newspapers for the previous century, and republishing them was not in itself an innovation. Abolitionists had already discovered that they could reconceptualize and effectively recapitalize such ads so that they no longer functioned as conventional notices of slaveholders seeking lost property, addressing other likeminded readers. Instead, if brought to a nonslaveholding readership, the same ads worked as exotic or troubling announcements, news from some other world. Incidental classified ads in one context became the instruments of pure moral suasion in another. The eighteenth-century British abolitionist Thomas Clarkson, in his groundbreaking *Abstract of the Evidence delivered before a select committee of the House of Commons in the years 1790 and 1791, on the part of the petitioners for the Abolition of the Slave Trade* excerpted runaway slave advertisements mentioning brands on face or hands, from a Jamaican newspaper.[6] In the United States, William Lloyd Garrison's Boston-based paper *The Liberator*, beginning with its sixth issue in 1831, reprinted ads for runaway slaves and slave auctions in a section called "Slavery Record." This reprinting turned the slaveholder's voice against himself.

When these ads were recontextualized in *The Liberator*, as Dan McKanan observes, "the slave owner became a witness against himself, testifying that violence was intrinsic to the property relation of slavery."[7] Soon, other journals took up the practice of using such "self-subverting quotation[s]."[8] Reprinting such ads was attractive because it removed abolitionist discourse from the abstract realm of rhetorical defense or opposition and crucially used the slaveholders' own words, spelled out in the brass-tacks language of commercial speech. The Grimké-Weld collaborative work, however, both subpoenaed that "testimony" and highlighted the role of data supplied by the thousand witnesses by omitting their own names from the work.[9] They shifted from a strategy that treated these ads as anecdotes or vignettes to one that reinterpreted them as the containers of data about the brutality of slavery. The marks, scars, and shackles that slaveholders noted as a means of identifying individual runaways became the individual, incremental indictments of slavery that might be systematically collected and analyzed. The ads were abstracted, their information pried loose and accumulated, aggregated en masse.

American Slavery As It Is was one in a multitude of projects that helped to create the modern concept of information, by isolating and recontextualizing data found in print. In his essay "Farewell to the Information Age," linguist Geoffrey Nunberg notes the shift in the nineteenth century from understanding *information* as the productive *result* of the process of being informed to a *substance* that could be morselized and extracted in isolated bits.[10] With its information abstraction, *American Slavery As It Is* became a model and a source for other abolitionist works like Harriet Beecher Stowe's 1853 *Key to Uncle Tom's Cabin* and William Goodell's 1853 *The American Slave Code in Theory and Practice*. But it is also a close ancestor of those forms of muckraking that have depended more on sifting public documents and putting their information into new juxtapositions rather than depending on going undercover or making secret materials public. From the 1950s to the 1970s the investigative journalist I. F. Stone, for example, scoured the *Congressional Record* and other government documents for many of his revelations. Such materials documented "contradictions in the official line, examples of bureaucratic and political mendacity, documentation of incursions on civil rights and liberties." His "use of government sources to document his findings was also a stratagem. Who would have believed this cantankerous-if-whimsical Marxist without all the documentation?"[11] Leaking and whistle-blowing may have a certain glamour, but Stone's journalism, like the Grimkés' and Weld's abolitionism, depended on something at once more subtle and more provocative of present concerns, not the opening of secrets but rather the painstaking extraction of already public information from the sources that have obscured it by dint of sheer proliferation. Don't think of Wikileaks, think of the power of search itself.

Both Grimkés had previously used their own testimony about slavery extensively in speaking and writing. In 1837, Weld had published *The Bible Against Slavery,* initially in the *Anti-Slavery Quarterly Magazine,* and then as a ninety-eight-page pamphlet. In it, he interpreted slavery in the Hebrew Bible as a form of paid service that could be stepped out of essentially at will, thus refuting claims that the Bible sanctioned chattel slavery as it was practiced in the United States. His biblical interpretation drew on another form of text mining, familiar to ministers: the concordance, essentially a keyword search through the text, providing context, in use since the thirteenth century. *American Slavery As It Is* importantly shifted the focus to the present when it took as its text the newspapers, along with testimony derived from questionnaires. It represented data mined from an enormous number of papers. Forty-five years later, Weld recalled:

> After the work was finished, we were curious to know how many newspapers had been examined. So we went up to our attic and took an inventory of bundles, as they were packed heap upon heap. When our count had reached *twenty thousand* newspapers, we said: "There, let that suffice." Though the book had in it many thousand facts thus authenticated by the slave-holders themselves, yet it contained but a tiny fraction of the nameless atrocities gathered from the papers examined.[12]

Weld noted that the sisters had "spent six months, averaging more than six hours a day"—the good daylight hours—"searching through thousands upon thousands of Southern newspapers, marking and cutting out facts of slave-holding disclosures for the book."[13]

With these large piles of papers, it became possible for the Grimkés and Weld to sort, categorize, and annotate what they found in the ads. The Grimkés used their expert knowledge as the Southern-raised daughters of a slaveholding family to identify some of the figures involved and to interpret the practices hinted at in the ads in newspapers that were, crucially, published by slaveholders. These ads were a weapon. In the words of Sarah Grimké, written to her friend Jane Smith as she worked on the book:

> Our present occupation . . . looking over southern papers, is calculated to help us . . . see the inside of that horrible system of oppression which is enfibred with the heart strings of the South. In the advertisements for runaways we detect the cruel whippings & shootings & brandings, practiced on the helpless slaves. Heartsickening as the details are, I am thankful that God in his providence has put into our hands these weapons prepared by the South herself, to destroy the fell monster.[14]

They mined the advertisements for information that they then sorted into categories such as "tortures, by iron collars, chains, fetters, handcuffs, &c.," "brandings, maimings, gun-shot wounds, &c.," and "Mutilation of Teeth."[15] They interpreted this data. Thus, for example, what might be the simple loss of a tooth in an era of bad dentistry, mentioned among other physical attributes in an ad for a runaway, is exposed as part of a scheme to identify slaves:

> Another method of *marking* slaves, is by drawing out or breaking off one or two *front teeth*—commonly the upper ones, as the mark would in that case be the more obvious. An instance of this kind the reader will recall in the testimony of Sarah M. Grimké . . . of which she had personal knowledge; being well acquainted both with the inhuman master . . . by whose order the brutal deed was done, and with the poor young girl whose mouth was thus barbarously mutilated, to furnish a convenient mark by which to describe her in case of her elopement, as she had frequently run away.[16]

These advocates thus took an undifferentiated pile of ads for runaway slaves, wherein dates and places were of primary importance, rendered in the neutral language of commerce, and transformed them into data about the routine and accepted torture of enslaved people.

Interpreted correctly, the ads yielded information on a horrifying spectrum of abuse, both of enslaved people's bodies and their spirits. Runaway ads documented the separation of families—evident in such items as "Runaway—my negro man, Frederick, about 20 years of age. He is no doubt near the plantation of G. W. Corprew, Esq. of Noxubbee county, Mississippi, as *his wife belongs to that gentleman, and he followed her from my residence.*"[17] The Grimké-Weld use of italics redirects the reader to Frederick and his wife's forced separation. Other advertisements illuminate the use of slaves in medical experiments. Beyond the ads, the Grimkés read other parts of the Southern press to take the pulse of the South. They clipped news stories that reported the jailing of enslaved children or extremely elderly people and news stories that celebrated the capture and punishment of runaways. The presence of these accounts unaccompanied by condemnation in public newspapers yielded evidence of "public opinion" in the South, a phrase *American Slavery As It Is* uses repeatedly. In a group of sections organized as responses to anticipated objections, the section demolishing the claim that "Public opinion is a protection to the slave" defines law as the distillation of public opinion, and sets forth the ways in which laws deprive slaves of rights and put them in danger.[18] It also uses newspaper ads as an index of public opinion, pointing out for example that the *New Orleans*

Bee with a large circulation among merchants, planters, and professional men, and thus "a fair index of the 'public opinion' of Louisiana," published ads for runaways identifying women by their whipping scars or physical deformities on parts of their bodies that would require that the women be stripped to show the marks, and thereby providing evidence that Southern public opinion did not object to this.[19]

Marking and reprocessing the newspapers allowed Angelina Grimké Weld and Sarah Grimké to compile the book's "many thousand facts thus authenticated by the slave-holders themselves."[20] Having sorted and categorized the data in the ads, the sisters and Weld offered various modes of sorting the same information in the body of the book. Testimonies of white former Southerners, or Northerners who had visited the South, or those still living in the South, for example, were presented as blocks of narrative. Then, some of the same material was extracted from the narratives and broken up into topics. While some topics, such as "Slaves suffer from hunger" were supplied mainly by brief extracts from narratives or questionnaires or other personal accounts, and others like "Punishments: Floggings" and "Punishments: Tortures" entirely from runaway ads, others, like "Clothing," drew together individual testimony, material from legal documents, and runaway ads. Those ads were central. The text explains: "We have . . . given to the testimony of the slaveholders themselves, under their own names, a precedence over that of all other witnesses." It follows with testimonies that back up these ads and "show, that the slaveholders who wrote the preceding advertisements, describing the work of their own hands, in branding with hot irons, maiming, mutilating, cropping, shooting, knocking out the teeth and eyes of their slaves, breaking their bones, etc., have manifested, *as far as they have gone* in the description, a commendable fidelity to truth." *American Slavery As It Is* moves recursively; each set of data is backed up by another level of evidence. The compilation of runaway ads supply solid evidence in the slaveholders' words, but further testimony confirms that specific slaveholders have committed these deeds. Personal testimony explains the analysis of runaway ads, and ads authenticate the testimony.

The book was made more usable to readers via a detailed table of contents and an index, which allowed for discontinuous, topical access. The table of contents breaks the sections down via headings and offers a nearly page-by-page digest, which forecasts and prepares the reader to be bombarded with horrifying particulars, as Stephen Browne notes.[21] Indexes generally serve as a bridge between author and reader, offering concepts, even if the author did not use a specific term directly. Indexes allow readers to access material from additional angles. One can use the *American Slavery As It Is* index to look up subjects such as "Lives of slaves unprotected, 155" and "chopping of slaves piecemeal, 93."[22] The index entries also editorialize: "plantations second only to hell, 114."[23] The individual and specific horrors were thus catalogued, sorted, and made accessible to be used as evidence in speechmaking or novel writing.

The index provided its users with tools for quick access to information, and enhanced the users' authority, and thus the authority of the book. Readers reported that they could use *American Slavery As It Is* to "stump" slaveholders—one said he related incidents of cruelty from the book, and when the slaveholders said they were lies, "he would pull Weld's volume from his pocket and give names, places, and dates from Southern papers."[24] The Grimké-Weld mode of reading the proslavery press so convinced readers of its reliability that they even felt confident substituting the Grimké-Weld readings for their own. As Louise Johnson discovered, and as Meredith McGill and Trish Loughran explore in relation to 1830s reprinting practices, Charles Dickens took the book up in his *American Notes* in 1842.[25] He quoted from it without attribution, recording specific ads that he lifted from *American Slavery As It Is* in his reports on his Southern travels as though he had come across the ads himself. In other words, he drew on a work compiled in New York and New Jersey from papers mailed from the South, to flesh out and provide detail for his own travels in the South.[26] Circulation and recirculation became a mode through which readers and travelers themselves came to understand the South and slavery. Writers like Stowe, Dickens, and the man who used the book to stump Southerners, relied on *American Slavery As It Is* for knowledge and details. But for ex-slaves speaking on the abolitionist circuit, it was a ready reference, containing information on laws having to do with slavery. Its "thousand [white] witnesses" carried authority that reflected back on the speaker's own statements. When Frederick Douglass read from it in an 1846 talk to English working people he thereby established that his own experiences and observations fit a larger pattern. He read aloud from the laws on slavery, recorded in *American Slavery As It Is* because "no better exposure of slavery can be made than is made by the laws of the states in which slavery exists. I prefer reading the laws to making any statement in confirmation of what I have said myself; for the slave holders cannot object to this testimony, since it is the calm, the cool, the deliberate enactment of their wisest heads, of their most clear-sighted, their own constituted representatives."[27] He extended the circuit of recirculation by recommending that his listeners read Dickens's *American Notes* for more information.

The representations of slavery in *American Slavery As It Is* were amplified and sent back out. The material recirculated back to the Southern newspapers, as well, and became a taunt, as Stowe reports: after the book was published, "a copy of it was sent, through the mail, to every editor from whose paper such advertisements had been taken, and to every individual of whom any facts had been narrated, with the passages which concerned them marked."[28] When Southern newspapers responded by attacking the volume, their attacks publicized *American Slavery As It Is*. Moreover, the book taught abolitionists and others a mode of reading the press. After drawing on *American Slavery As It Is* in *Uncle Tom's Cabin*, Harriet Beecher Stowe then used it in composing *The Key to Uncle Tom's Cabin*, quoting, for example, the "testimony" given by "an unimpeachable witness, Miss Sarah M. Grimké" of the iron collar being used on "[a] handsome mulatto woman."[29] Stowe's *Key*, according to Thomas Leonard, "recycled Weld-Grimké clippings and added her own from more than 200 southern papers to support her novel."[30] Stowe notes of *American Slavery As It Is* in her *Key* that "the papers from which these facts were copied were preserved and put on file in a public place [the office of the American Anti-Slavery Society], where they remained for some years, for the information of the curious."[31]

Slaveholders objected that the information represented in these sources reflected atypical situations, but the sheer number of newspapers comprising the Grimké-Weld database provided a strong refutation to that argument. In fact, the slaveholders had a point: in any single paper that *American Slavery As It Is* quotes, there might be only one such ad, and, in fact, the Grimkés must have looked at some papers that yielded nothing. Moreover, someone taking up the invitation to visit the office of the American Anti-Slavery Society to read the newspapers they had used might find the ads in very different form. The Grimkés and Weld sharply edited the ads to eliminate most of the identifying information about the ex-slaves, ensuring that the notices they had converted to raw material for their antislavery arguments did not inadvertently revert to their original function and provide information that would lead to recapture. They also trimmed to highlight the points they wished to focus on—the separation of families and not the escapee's "rather sulky appearance," for example.[32] Trimming also concentrated the information, making it easier to compile. When a single ad was placed alongside dozens of other similar advertisements, the information became a data point in a wide-ranging representation of a common practice. They noted, too, that they had selected the runaway ads that they included from many others that they could have quoted, and that these were representative of a larger whole: "Scores of such advertisements are in Southern papers now on our table. We will furnish the reader with a dozen or two."[33] It was the work of trimming, sifting, and aggregating the material that recreated it as a database and not just a collection of anecdotes. This work allowed for its recontextualization and analysis.

The Grimké-Weld household's project of mining the newspapers was made possible by access to large piles of Southern newspapers. Where did they get the papers? They could have looked to the reading rooms and public libraries of New York City, locations crucial to spreading ideas and information. Some reading rooms were political, offering print resources in support of a cause. For example, in the late 1830s there were at least two antislavery reading rooms in New York City, both run by antislavery newspapers, making use of the newspapers and magazines they received in their exchanges. Like other newspapers, abolition papers exchanged free copies with other publications and were allowed by the U.S. Post Office to mail exchange copies without cost. The antislavery papers received copies of pamphlets, as well, which they made available to visitors who paid a yearly or weekly fee. Both of the New York reading rooms framed their projects as offering resources to the black community.

One New York reading room was run by David Ruggles, an African-American activist with the New York Committee of Vigilance, which watched out for and fought against kidnappers and slave catchers, and editor of the *Mirror of Liberty*, whose office was located at Lispenard and Church Streets. In May 1838 he complained that black men were excluded from "Reading Rooms, popular lectures, and all places of literary attractions and general improvement," and announced that he had opened a reading room at the Committee's office, which was also his home, offering "access to the principal daily and leading anti-slavery papers, and other popular periodicals of the day."[34] His reading room offered access to these papers for a fee—from $2.75 per year to 6½ cents per week, waived for "strangers visiting this city"—including the many fugitive slaves Ruggles was in contact with.[35] Another antislavery reading room soon opened less than half a mile away, sponsored by the weekly *Colored American*, a black-owned newspaper. In January 1839, the paper announced that it planned to offer to "friends and subscribers" a place to read the other papers they received in their exchanges. While those papers would have included the abolitionist press, the *Colored American* exchanged with others as well. It announced, "Our Files are well filled with the principal Foreign and Domestic papers—Religious, Moral, Literary and Political."[36]

While these antislavery reading rooms—and others around the country, like the one Frederick Douglass ran in Rochester—were valuable as sites for following the movement, spreading knowledge of events and tactics, and possibly for education and self-improvement as their prospectuses proclaimed, they were not extensive repositories.

The Grimkés and Weld surely did not draw their wealth of evidence from them. Other newspapers, too, offered visiting readers some form of access to their exchange papers, but even those would not have been sufficient for the purposes of the Grimkés and Weld, especially since most newspapers were partisan and more likely to exchange with like-minded publications, and therefore have only such newspapers on hand. A different kind of home-based exchange became relevant here for the thorough and extensive coverage that gave *American Slavery As It Is* its powerful evidentiary status.

According to Trish Loughran, Theodore Weld commuted daily from his home in Fort Lee, New Jersey, to his office on Nassau Street in Manhattan, where he purchased "in bulk" newspapers that were to be "sold for waste when their newsworthiness expired," and brought them home to Fort Lee, where "the Grimkés performed 'their daily researches' at the kitchen table."[37] Forty years after the fact, Theodore Weld reported that he had purchased the more than twenty thousand papers comprising his database—all the "papers published in the Southern States and Territories," somewhere between six months' and two years' worth—from a reading room that he recalled as the New York Commercial Reading Room.[38] This was almost certainly Gilpin's Merchants' Exchange Reading Room, a large room located inside the New York Stock Exchange which received hundreds of newspapers from around the country and the world.[39] The runaway ads and other data which Weld and the Grimkés drew from these pages originated in the same web of commerce that merchants were deeply interested in. Just as the words of the slaveholders could be turned against themselves, institutions like the Exchange reading room that commerce depended on could be turned from their tasks of commerce and used against themselves.

Ruggles had complained of the whites-only policy of reading rooms like this one. As well, like most public spaces associated with commerce, the room probably did not admit women; an 1863 engraving of the interior shows an all-white, all-male clientele.[40] The collaboration between Theodore Weld and the Grimké sisters, then, allowed the sisters access to an immense lode of data from which they would otherwise have been barred; their labor and expertise made the processing and reading of this data possible. Not only had their collective project imagined data by dint of its displacement from its original Southern contexts, their project had also depended on an additional displacement of those same sources from the commercial to the domestic sphere.

The extraordinary repurposing, reuse, and, most important, reconceptualizing and new juxtapositioning of media represented by *American Slavery As It Is* entailed a complex negotiation between modes of access to media, expertise, and the imagination to see that Southern newspapers not only could be made to speak against themselves, but also could be picked through, tagged, and sorted to support a new mode of understanding. That new mode of understanding might be called informatic, though informatics—like computers—of course lay many years in the future. Weld and the Grimkés arrived at an informatic sensibility out of the growing sense of urgency that abolitionists felt—the sense that simply softening the hearts of slaveholders was ineffectual and that hard facts were needed—which impelled them to turn to a new way of working. Like present-day academic researchers who pick through databases for particular uses of words, for authors' names, or for fragments of poetry to place them into new contexts that will yield new interpretative possibilities, Sarah and Angelina Grimké and Theodore Weld reconceived of ads and articles in proslavery papers as alienable bits—as content—that could be broken free of context and aggregated, strung along different threads to yield a damning portrait of slavery written in the slaveholders' own words.

Notes

1. *American Slavery As It Is: Testimony of a Thousand Witnesses* (New York: Arno, [1839] 1968). On authorial attribution, see note 9 below. Critics have also argued that the book—addressed to readers who might never have seen black people—popularized a focus on enslaved people's inhuman suffering, which made them seem less human.

2. See Trish Loughran, *The Republic in Print: Print Culture in the Age of U.S. Nation Building, 1770–1870* (New York: Columbia University Press, 2007), 355, for details of how the testimonies of first-hand witnesses to slavery's abuses were gathered in response to a widely distributed "personalized circular letter" reproduced using the new technology of lithography. Harriet Beecher Stowe also reported these procedures in *A Key to Uncle Tom's Cabin: Presenting the Original Facts and Documents Upon Which the Story Is Founded* (Port Washington, NY: Kennikat, [1853] 1968), 21.

3. Jean Fagan Yellin, "Doing It Herself: *Uncle Tom's Cabin* and Woman's Role in the Slavery Crisis," in *New Essays on Uncle Tom's Cabin,* ed. Eric J. Sundquist (Cambridge, UK: Cambridge University Press, 1986), 90.

4. Benjamin P. Thomas, *Theodore Weld: Crusader for Freedom* (New Brunswick, NJ: Rutgers University Press, 1950), 172.

5. Gilbert H. Barnes and Dwight L. Dumond, eds., *Letters of Theodore Dwight Weld, Angelina Grimké Weld, and Sarah Grimké, 1822–1844*, vol. 2 (Gloucester, MA: Smith, 1965), 789.

6. Adam Hochschild, *Bury the Chains: Prophets and Rebels in the Fight to Free an Empire's Slaves* (Boston: Houghton Mifflin, 2005), 197.

7. See Augusta Rohrbach, "'Truth Stronger and Stranger than Fiction': Reexamining William Lloyd Garrison's *The Liberator,*" *Truth Stranger Than Fiction: Race, Realism, and the U.S. Literary Marketplace* (New York: Palgrave, 2002), 2, for an examination of the relationship of *The Liberator* to "liberal capitalism and moral suasion."

8. Dan McKanan, *Identifying the Image of God: Radical Christians and Nonviolent Power in the Antebellum United States* (New York: Oxford University Press, 2007), 135.

9. Although writers like Spender have criticized Theodore Weld for failing to share authorial credit for *American Slavery* with his wife and sister-in-law, and he did sign the circular requesting information, his name does not actually appear on the 1839 edition as the book's author. Authorial credit seems to have been assigned to Weld later. Moreover, the unsigned Introduction, which convenes the book's readers as the jury that will weigh the witnesses' testimony, explicitly empanels both women and men on the jury. Dale Spender, *Women of Ideas and What Men Have Done to Them: From Aphra Behn to Adrienne Rich* (London: Routledge, 1982), 166.

10. Geoffrey Nunberg, "Farewell to the Information Age," in *The Future of the Book*, ed. Geoffrey Nunberg (Berkeley: University of California Press, 1996), 103–138.

11. Victor Navasky, "I. F. Stone," *The Nation* 277 (July 21, 2003): 17.

12. Quoted in Catherine H. Birney, *The Grimké Sisters: Sarah and Angelina Grimké, the First American Women Advocates of Abolition and Woman's Rights* (Boston: Lee and Shepard, 1885), 258–259.

13. Quoted in Birney, *The Grimké Sisters*, 258.

14. January 24, 1839, quoted in Thomas, *Theodore Weld*, 168–169.

15. *American Slavery*, 72, 77, 83.

16. Ibid., 83.

17. Ibid., 164.

18. Ibid., 143.

19. Ibid., 154.

20. Birney, *The Grimké Sisters*, 259.

21. Stephen Browne, "'Like Gory Spectres': Representing Evil in Theodore Weld's 'American Slavery As It Is,'" *Quarterly Journal of Speech* 80 (August 1994): 277–292.

22. *American Slavery*, 214, 212.

23. Ibid., 217.

24. Thomas, *Theodore Weld*, 172.

25. Louise H. Johnson, "The Source of the Chapter on Slavery in Dickens's American Notes," *American Literature* 14 (January 1943): 427–430; Trish Loughran, *The Republic in Print: Print Culture in the Age of U.S. Nation Building, 1770–1870* (New York: Columbia University Press, 2007); Meredith L. McGill, *American Literature and the Culture of Reprinting, 1834–1853* (Philadelphia: University of Pennsylvania Press, 2003).

26. *American Slavery*, 127–128.

27. Frederick Douglass, *American Slavery: Report of a Public Meeting Held at Finsbury Chapel, Moorfields, to Receive Frederick Douglass, the American Slave, on Friday, May 22, 1846* (London: C. B. Christian, 1846).

28. Stowe, *Key*, 21.

29. Ibid., 89; see also 21 and 90.

30. Thomas C. Leonard, "Anti Slavery, Civil Rights, and Incendiary Material," in *Media and Revolution: Comparative Perspectives*, ed. Jeremy D. Popkin (Louisville: University Press of Kentucky, 1995), 121.

31. Stowe, *Key*, 21.

32. *American Slavery*, 166, includes a four-line item "from the 'Richmond (Va.) Whig,' June 30, 1837. 'Ranaway, my man Peter.—He has a *sister* and *mother* in New Kent, and a *wife* about fifteen or eighteen miles above Richmond, at or about Taylorsville. Theo. A. Lacy.'" This emphasis, of course, did not appear in the much longer advertisement as it was actually printed, which also noted the man's skin color and build, sulky appearance, and what he was wearing ("Notice").

33. *American Slavery*, 83.

34. "Circular," par. 2, *Colored American*, June 16, 1838: n.p. Accessible Archives, www.accessible.com (accessed April 30, 2012).

35. Graham Russell Hodges, *David Ruggles: A Radical Black Abolitionist and the Underground Railroad in New York City* (Chapel Hill: University of North Carolina Press, 2010), 135.

36. "READING ROOM," *Colored American* (January 12, 1839): n.p. Accessible Archives, www.accessible.com (accessed April 30, 2012). See also "A Reading Room," *Colored American* (February 10, 1838): n.p. Accessible Archives, www.accessible.com (accessed April 30, 2012).

37. Loughran, *Republic in Print*, 355, 356, 357.

38. Birney, *The Grimké Sisters*, 258–259.

39. Other than Weld's mention of it, I have found no references to the New York Commercial Reading Room. The size of Gilpin's Exchange, its proximity to the office of the American

Anti-Slavery Society offices, and the fact that the *Emancipator* had used its resources shortly before the Grimké-Weld project began suggests that forty years after the fact, Weld simply got the name of the room wrong, possibly mixing it up with the name of one or more similar rooms in other cities ("Reading Room Gleanings," *Emancipator* 42.163 [February 15, 1838]: col. A). In 1834, Gilpin's cost seven dollars a year to use, or, for "strangers," seventy-five cents per month; see Edwin Williams, *New York As It Is, in 1834; and Citizens' Advertising Directory Containing a General Description of the City and Environs* (New York: Disternall, 1834), 159.

40. An engraving showing the interior of the "Reading Room of the Merchants Exchange and Newsroom" appeared in *Frank Leslie's Illustrated Newspaper*, May 30, 1863, 147. It is on the Mr. Lincoln and New York website (Lincoln Institute): http://www.mrlincolnandnewyork.org (accessed January 28, 2012).

6 Paper as Passion: Niklas Luhmann and His Card Index

Markus Krajewski
translated by Charles Marcrum II

Card indexes can do anything!
—*Das System, Zeitschrift für Organisation*, Book 1, January 1928

Hegel's absolute spirit is a hidden slip box.
—Friedrich Kittler

Prologue

Somewhere in Stuttgart, 1785: Still in high school, a fifteen-year-old reader begins to write on loose sheets of paper with order, diligence, and discretion: "In his reading, he approached works in the following way: everything that seemed noteworthy to him— and what didn't!—he wrote on a single sheet, which he labeled above with the general heading under which the particular content should be subsumed. In the middle of the upper edge, he then wrote the keyword of the article in large letters, frequently in Fraktur. He organized the sheets themselves again according to the alphabet, and due to this simple mechanism, he was always ready to use his excerpts at any moment."[1] With each of his alphabetized notes, the young reader established a new address that would henceforth constitute the site for the concepts upon which his future activities as philosopher and scholar would be based.

Whether Hegel's famous slip box (*Zettelkasten*, a sort of card index) remains undiscovered in the estate of the Prussian Cultural Heritage Foundation—and thus represents an unsolved problem of address for research questions[2]—or is actually lost, we can nevertheless gain an impression of Hegel's practice of excerpting. Not only do his papers include handwritten cards that suggest the desired path of the student (with keywords such as "academy," "pedagogy," "way of teaching," and "erudition of the Egyptians"), manuscripts from Hegel's Stuttgart period contain excerpts, such as one on the "Path to Happiness in the Great World," copied verbatim from Johann Georg Zimmermann's *Über die Einsamkeit*. Moreover, in the same manuscript—in addition to this excerpt— there are other passages on the theme of bliss, pulled from earlier readings in other texts. Is a systematic ordering of the notes preferred here to an alphabetical grouping? In either case, the early order of the slip box reveals the ways "that Hegel's mind occupied itself systematically with general, important problems."[3]

Somewhere in Bielefeld-Oerlinghausen, 1997: A scholar and philosopher, now in his seventies, observes himself, absorbed in reading: "When I read a book, I do it in the following way: I always have a note card on hand, on which I note the ideas of certain pages. On the back side, I record the bibliographical information. When I have read through the book, then I go through these notes and consider what could be useful for notes that have already been written. Thus, I always read with an eye toward the notability of books."[4] The professor emeritus still tends his significant and legendary collection of notes and handwritten cards, in that he links the comments and ideas from his readings with the previous concepts in the form of "cards in Octavo format,"[5] complete with abbreviations and keywords, which await connections in the registry of his index. "What does one do with what has been written?" Niklas Luhmann asks: "To be sure, one will initially produce mostly waste. But we have been raised such that we expect something useful from our activities or otherwise quickly lose heart. Thus, one should consider whether and how to process the notes so that they are available for later access, or at least provide such a comforting illusion."[6]

It is worth pursuing this consolation, following the question of the materiality on which the system of note taking in Bielefeld was based after 1951, which library-oriented and informatic techniques of data processing and regeneration this system obeys, in contrast to the system—Hegel's—of Stuttgart, Tübingen, Bern, Heidelberg, Frankfurt, Nürnberg, Jena, and Berlin, 1785–1831. Such an inquiry ultimately allows a look at the self-representation of the system and its production aesthetic, and at the internal communications situation that figures, in turn, as the basis of the theory of self-referential systems. In the process, the history between 1785 and 1951 from the side of learned discourse will be deliberately suppressed, with only occasional digressions about innovations in library and notation technologies surfacing between the two dates, in order to clarify the particularities of the Bielefeld 1951ff. notation system against an implicit historical template, for, "In contrast to other names, the name Hegel appears in his work not only as an historical object through which his own theory was tested, but also as a competitor against whose accomplishments it was measured."[7]

The materiality of systems theory—also known by its bureaucratic classification as "Theory of Society; Duration: 30 years; Costs: None"[8]—draws its productive power, persistent for three decades, from the generative pairing of man and machine. When Niklas Luhmann decided in 1951, toward the end of his legal studies, to no longer gather loose sheets into portfolios, as Goethe once did,[9] but rather to take up work on a slip box, just like his implicit benchmark Hegel, the position of the Other became occupied by a paper machine. The generally connoted opposition of man . . . machine[10] loses its validity in systems-theoretical terminology. Instead, both systems partaking in the communication ("that we regard ourselves as systems will surprise no one"[11]), the "psychological" and the "system of notes,"[12] form a constellation that is shaped by the term "partnership." But before that dispositive undergoes closer inspection in the second part of this chapter by means of a voyeuristic look at this togetherness, we must first examine the systemic (and systematic?) construction of order.

The Scholar Machine

He excerpted continually, and everything that he read went from one book next to his head and into another.
—Georg Christoph Lichtenberg, *Sudelbücher*, Book G 181, 1779

Led by the apparently frequent question regarding the criteria according to which his slip box was constructed, Luhmann willingly granted a glimpse into the architecture of the system and its features, commenting in an interview: "By the way, many people have come here to see that."[13] The writing tool became an object of desire, especially for young academics seeking to add a carefully planned card index to their carefully planned careers: "After all, Fred wants to be a professor."[14] The precise directions for recreating the slip box appear in 1981 in a special volume for Elisabeth Noelle-Neumann as "a piece of empirical social research,"[15] which then nevertheless delivers a precise, theory-saturated description of how a sociological supertheory is to be encoded.

According to this, the arrangement consists of "wooden boxes with drawers that pull out in the front, and cards in octavo format" (= DIN A5). One needs to be mindful of space-saving facilities, in order to still be able to comfortably handle the sometimes very large paper collection after decades of care: "Because I need space. Not for my stomach (an article in which I can't fit much), but rather for slip boxes, portfolios, and above all the library table, positioned around me very precisely in quadrants, in such a way that I can still reach it, bent forward impetuously—I have long arms!"[16] Thus, another theorist of the slip box defends the place of the writing instrument, not coincidentally at a time when the triumphant march of the personal computer into the living rooms of the world was about to begin. Despite the librarian card-theoretical recommendation of only using cardboard or strong paper as a bearer of information,[17] Luhmann relies on plain typewriter paper for spatial economy, which can quickly lead, however, to the deterioration of the medium with frequent browsing. While the administrative scientist Luhmann ignores the librarian's dictum in his consideration of the proper paper for the project out of spatial concerns, DIN 1504, which, apart from the International Library Format, only allows DIN A 6 and DIN A 7 for "literature cards,"[18] regrettably goes unused.

As in the Stuttgart System 1785ff., each literature card initially represents one keyword, which can sometimes be greatly differentiated and expanded. As identification, the cards of unreinforced, normal paper (in order to save space) bear a unique key in the upper-left corner. "Hagen: They have their own shorthand. Luhmann: Yes. Each card has a concrete number that is never changed. H: And that's shorthand? L: Yes, yes. H: Does the number have some kind of meaning? L: No."[19] Not least for mnemotechnical reasons, the shorthand begins numerically, followed by a slash, to which another number is attached, which is then either incremented in another entry or branched "internally" with a new enumerator in the form of a lowercase letter, for example 43/12a or 43/13. The shorthand serves as a concrete address for each card, which may thus not be rearranged into a thematically centered system or an alphabetical catalogue. The process of simply adding each new entry to the end, so long as there is no "internal" connection, is familiar to every librarian under the term *numerus currens*, and it reveals its strength in the simplicity of locating cards by means of their alphanumeric shorthand: if there is a reference within a new note to a term that has already been recorded, it is sufficient to record the shorthand for the note that is to be referenced next to the newly recorded word. This possibility of massive referencing guaranteed the "buildup of high complexity in the slip box."[20]

However, the cards did not remain assembled in their chronology, in other words, 43/12, 43/13, 43/14 . . . because it is permissible to branch a term internally. For example, the card for the keyword "differentiation" undergoes a decisive and distinctive explanation. Thus, the card for the keyword "differentiation, hierarchical" receives not the shorthand 43/77, but rather 43/76a, including the reference to the card for "hierarchy." As an effect of this "internal branching capability,"[21] individual "clusters" of terms form, which assume central importance for the theory (and vice versa).

An alphabetical register, in which each new card—whether branched internally from an existing "cluster" or added to the end—appears with an appointed shorthand, serves as a search engine that allows access to the desired term. By means of this register, which gradually writes itself in bound form—like the hardback catalogue of the nineteenth century—one gains access to a term, on whose card commentary refers to the connections that form in different places within the slip box, and so on, just as on the other cards. Thanks to this possibility for connections, once picked up and gently led by the network of references, the structure of the text to be written appears in its early form. By leafing broadly through the cards, they link together one by one, with their sequence anticipating the loosely linked sheets of the later text: "This technique also explains why I don't think at all linearly and have trouble finding the right sequence of chapters when writing a book, because indeed every chapter must reappear in every other."[22] All that the emerging text lacks is a means of bridging the gap between the selected comments from the respective sources in card form, which are largely filled through "rephrasing"[23] during the reading of books.

Numerous cards serve as building blocks of the text being composed, which are worth transferring from the preselected contingency into the order of a still-one-dimensional textual structure. The decision that arises during the collecting/browsing process to pursue one reference and not another, and to prefer a perhaps esoteric card, which then offers entirely different connective possibilities, and build it into the string of terms assures calculated coincidence a secure position in the combinatorial calculus. Thanks to this, the plan for drafting a text undergoes surprising changes. In the Bielefeld 1951ff. notation system, the slip box becomes a combination machine, which not only answers the questions asked of it with some remembered reading, but also offers a list of connections, in order to connect the following argumentation with its terminological and bibliographical resources. Thus, it is important to distinguish between two kinds of literature cards: cards that contain comments, excerpts, and lines of reasoning on a topic versus cards that solely present bibliographical information. The first variety, which represents the largest portion in the Bielefeld 1951ff. recording system, consists of nothing more than a classic thesaurus, a treasure of theory with no alphabetical order, which can contain not only brief explanations for the desired terms, but also, sometimes, long collections of material. The latter serve as sources, on which the contents of the first variety gorge themselves, not—as in the Stuttgart System 1785ff. and the *Phenomenology of Spirit*, its first great product[24]—to suppress, but rather to anchor the inspiration of the thought in the "sea of scholarship."[25] The medium that first made this combinatorial quotation possible is not the quill or the typewriter, but rather the "paper machine," and not in a trivial sense. Luhmann: "In this respect, I work like a computer, which can also be creative in the sense that it produces new results which were unforeseeable through the combination of data."[26] Thus, the manual yet easily automated paper processing with the heuristic label Bielefeld 1951ff. comes remarkably close to another process already established in 1936: the paper machine.[27] It does not have individual cards, but rather an infinitely long paper band, as well as strictly defined work instructions[28] and a read/write head,[29] and thus it becomes a universal machine, in order to merge entirely into this breach in the history of data, office, and paper processing introduced in 1937.

Potential possibilities for improving the Bielefeld 1951ff. recording system emerge almost inevitably with the steady triumphant progress of the now universal paper machine of 1937, which produces new computer generations and calculating speeds on a quarterly basis. Logically, an electronic slip box allows one faster access to random terms and likewise, in combination with logical connections, to never overlook—or forget—character strings in the electronic resources. Thanks to hypertext, the idea for which goes back to Vannevar Bush's thought-expanding machine Memex from 1945,[30] the formerly tediously annotated references can be traced and (automatically) connected with an equally time-optimized strategy of click and rush.[31]

However, even if Luhmann's method follows a clear algorithm, and he functions in a certain sense as a computer, this is still a long way from a digital notebook or laptop. For example, although Hegel's slip box in handy luggage format joined him for every journey and all seven migrations to Berlin,[32] the many square meters of Luhmann's wooden boxes prevent unlimited mobility and thus the possibility of accessing written memory at all times. "H: [T]his card architecture and these box dimensions are massive, aren't they? L: It is somewhat comprehensive, yes. H: A few meters. L: Yes, yes. H: . . . and is the basis, so to speak, of your work. L: Yes. H: Without which . . . If one were to take that from you, then it would be difficult. L: Yes, then it would be difficult."[33] The communicative partner, thanks to whose indispensible help the theory achieves its legendary productivity, remains in its usual place, in order to await there the questions that are directed toward its wood. "Fred admires this systems theorist from Bielefeld (from Biiiielefeld, I always say, while raising my eyebrows high)."[34]

With this attempt to shift a small city in northeastern Westphalia to Silicon Valley and to suggest the consequently chance-driven combinatorics of the basic systems-theoretical code, the materiality of the strange system has been revealed. Our focus may now be directed toward this machine and the form that the connections take, making the slip box and its operator the fundamental and therefore privileged system of systems theory.

Partnership Agreement

I am not talking about an assistant or about "psychic systems"[35] called upon for help. For the institution "university," Bielefeld 1951ff. maintains no personal external reference ("Costs: none"). Rather, the systems theory writes (itself) with the help of a communicative collective of equal partners, which consists of "me and my slip box."[36] The production aesthetic appears as a dispositive, with its constituents of man and machine politely facing each other, figuring under the term "partnership."

It remains to be asked why it is acceptable to speak of the paper-machined slip box as a "communicative partner"? What justifies placing both parties onto one dialogical plane and thus flattening significant differences in the (naively) assumed communicative competence of both sides? How does the anthropomorphization of the wooden box take place?[37]

Intrinsic to systems theory, an initial answer lies in the construction of the term "communication" itself. As a "system of notes," the wooden box has the possibility of accepting or rejecting communication in the form of inquiries directed toward it. As Bielefeld 1951ff. assumes that the slip box is a system, and further, since systems theory requires that there be systems,[38] the fundamental constitution of such a system applies to Bielefeld 1951ff. as a social system, formed from the slip box as a notation system and the "I" as read/write head: "To the question of what constitutes social systems, we therefore give the double answer: communications and their attribution as action."[39] Fifteen years later, in the question of what ultimately is the basis for the constitution of a social system, action or communication, the latter has proven more amenable as a theoretical design. Communication "is the smallest possible unit of a social system, namely that unit to which communication can still react through communication."[40]

How the fundamental term "communication" is conditioned and the implications that arise from it have already been adequately explained elsewhere.[41] In this context, only the modular structure of communication remains to be emphasized, only ever capable of joining with its own kind as the smallest building block of a social system. Following a communicative prompt, like the question about the card dispositive—how is a form to be created?—is the reaction of the box, which either accepts or denies the communication, in other words, answers with its own communicative act that either declines by showing its blank space or offers new information about the inquiry—that is, by coming up with an excerpt. Once initiated, the slip box continually produces new connections and cross-references, which present the far-reaching, sometimes meandering answer to the inquiry: "Communication is . . . autopoietic insofar as it can only be produced in a recursive relationship to other communications, that is to say, only in a network, to the reproduction of which each individual communication contributes."[42]

Apart from the argument of the language common to both sides, why can one speak of a communicative process within Bielefeld 1951ff.? First, inasmuch as the slip box has a critical mass of entries as well as a certain number of cross-references, it offers the foundation for a particular form of communication, for its own poetological process of knowledge production, which can help its users to entirely unforeseen insight. If slip box practitioners like Luhmann assume that they find an equally valuable and stimulating communicative partner in the wood and paper apparatus, this thesis, in turn, reaches back to a constellation already described in 1805 by Heinrich von Kleist in his fascinating analysis of the "Midwifery of Thought": "If you want to know something and cannot find it through meditation, I advise you, my dear, clever friend, to speak about it with the next acquaintance who bumps into you."[43] The positive tension that such a conversation immediately elicits through the expectations of the Other obliges one to produce new thought in the conversation. The idea develops during speech. There, the sheer availability of such a counterpart, who must do nothing further (i.e., offer additional stimulus through keen contradiction of the speaker) is already enough; "There is a special source of excitement, for him who speaks, in the human face across from him; and a gaze which already announces a half-expressed thought to be understood often gives expression to the entire other half."[44] Thus, Kleist's basic idea is that communicative partners, in order to gain clarity about what they intend to express, need a silent catalyst for insight. What does the Other provide through its mere presence? In a word that Kleist borrows from Kant: a "midwifery of thought."[45] Without such an Other, a kind of "intellectual bankruptcy" threatens; however, with such a communicative partner, abundance beckons. It is not without reason that analogies are frequently drawn between the collections of written material and excerpts and the so-called card banks of the seventeenth century.

According to Kleist, the "human face" already serves as a sufficient source of inspiration, in which "a gaze which already announces a half-expressed thought to be understood" is sufficient, because such a gaze also "often gives expression to the entire other half." Now, one could say that the gaze toward wooden drawers usually supplies rather meager inspiration. However, what if we replace "human face" in this decisive quote with "interface," that is, the interface between man and apparatus? And if we replace the simple word "gaze" with the equally insignificant "touch"? For it is precisely the touching of the cards in their drawers, through interplay with this interface, which makes the silent Other talk—even more than Kleist postulated. The slip box offers an interface that is more than just a pleasant sight, in that the apparatus delivers those keywords that stimulate the protagonist to further thought production in response to a simple touch. The previously silent Other becomes a proper interlocutor. The finely branched network of connections guarantees that the keywords that are subsequently exchanged appear by no means haphazardly. For over the course of the operation, these gradually cultivate a "kind of second memory"[46] within the apparatus. And this second memory gains a certain independence when it intervenes in the stream of thought of the reasoning Other even more than Kleist proposed.

From this self-contained device for knowledge production in the form of an independently thinking slip box, it is merely a small step to imagining the agents of knowledge as networked and envisioning an organization of texts that renders all human intervention superfluous. Thus, the sensitive seismographer of avant-garde developments, Walter Benjamin, logically conceived of this scenario in 1928, of communication with card indices rather than books: "And even today, as the current scientific method teaches us, the book is an archaic intermediate between two different card index systems. For everything substantial is found in the slip box of the researcher who wrote it and the scholar who studies in it, assimilated into its own card index."[47] Thus, the intermediary subject seems to be overcome, if slip boxes now connect directly with slip boxes, in order to establish unadulterated data streams that transfer with less error.

One of the basic philological precepts states that "the text knows more than the author." One could easily carry this dictum over to the relationship between the slip box and its user. In their relay potential, the text fragments held ready by the apparatus offer far more connections than the inquirer is aware of at any moment. Hence, the interface offers an abundance of possible connections; it delivers the action potential of new argumentation. The slip box knows more than the author, in that it hides the conditions of knowledge and helps to catalyze future thoughts through contacts with its "interface."

For, to once again invoke Kleist, "It is not we who know, it is first and foremost a certain condition of ours which knows."[48] And it is precisely these possible conditions that the apparatus holds unwaveringly. Through its elements, prefabricated for connectivity, it always offers a configuration of potential states of knowledge, which are only realized, that is to say retrieved, by the user through certain combinations at a given time.

Consequently, the dialogical moment in the poetics of systems theory, put more concretely, the wooden partner within the collective, merits a status that is hard to overstate, which for its part bids farewell to that production aesthetic that shone for so long under the notion of Genius. As Goethe put it,

> In my nightly vigils the same thing happened: I therefore often wished, like one of my predecessors, to get me a leather jerkin made, and to accustom myself to write in the dark, so as to be able to fix down at once all such unpremeditated effusions. So frequently had it happened, that, after composing a little piece in my head, I could not recall it, that I would now hurry to the desk, and, at one standing, write off the poem from beginning to end; and, as I could not spare time to adjust my paper, however obliquely it might lie, the lines often crossed it diagonally. In such a mood I liked best to get hold of a lead pencil, because I could write most readily with it; whereas the scratching and spluttering of the pen would sometimes wake me from my somnambular poetizing, confuse me, and stifle a little conception in its birth.[49]

Constant doubts arise regarding glorifying self-descriptions of creative power, not only against the background of Bielefeld 1951ff., which certainly could have been ready in its materiality in Weimar "around 1800." Nevertheless, the precautionary withdrawal of that progressive tendency remains, remarkably, in the texts that could evoke justified skepticism regarding the myth of the genius, because it makes its own development transparent. It is similarly obscure in Thüringen, or more precisely: in Jena in 1806, as a certain young philosophy professor and master of Stuttgart 1785ff. sends the manuscript of the *Phenomenology of Spirit*, revised with complete comments, to his publisher in Bamberg ahead of the marauding French troops—as the story goes—through the last messenger still capable of leaving the city. No time for footnotes? Or did he destroy the remaining notes just in time? "In order to make all other philosophical textbooks superfluous, to make itself indispensable, the new textbook simply deleted its addresses. With the result that footnote-less Hegel students famously just read more Hegel and their master, in order to have any data at all for processing, had to invent 'the human' as a new data source."[50]

The Secret of the System

Though routine represents one of the stabilizing elements of a partnership, and recurrent actions and consistent processes are well rehearsed in daily experiences, the cooperative collective Bielefeld 1951ff. is particularly motivated by the continuous ability of the slip box to stimulate the curiosity of the inquirer through its accrued abundance and broad resources. Two effects contribute to the tracing of new relationships, that is to say, different and unintended ways of reading, following the reference structure of the entries: the surprise of stumbling upon an aspect not previously considered thanks to a reference, and the coincidence of this hint appearing precisely there and not elsewhere. In turn, from the contingency of the coincidental—provided sufficient space for development—grows the chapter problem mentioned previously, the question of how the issue to be related should be organized. More decisive, however, is the ability of the partner to surprise the inquirer.

By means of headwords and shorthand, one can refer precisely from any point in the slip box to another. In contrast to the book, with its rigid binding and its unavoidable format standards, each card represents a distinct, expandable unit of information, an extensible, elementary datum that is easily referenced. For each card bears a unique address thanks to its position in the order or in the form of a shorthand that can then be referred to by other cards: "Each note is just an element which first gains its quality from the network of references and cross-references in the system."[51] By means of these cross-references, the user can now trace new relationships, that is to say, previously unintended modes of reading, following the reference structure of the entries: "From given input, the slip box produces combinatorial possibilities which had never been planned, intended or conceived of as such."[52] Thus surprise arises, after stumbling across an aspect not previously considered thanks to an unexpected reference.

Accordingly, how does one succeed in endowing the slip box with this ability for surprise, supported by an indispensible abundance of information? In a word: time, with which complex structures will develop without outside assistance. These will arise so long as it is guaranteed that the user injects information in the form of small textual building blocks, facts, thought fragments, longer excerpts all the way to completely predetermined lines of reasoning with sufficient persistence, and moreover that these be tied into the existing referential structure. The potential to surprise owes more to the informed slip box than to a reading effect. As the assembled notes remind one of the thoughts thought through the duration of the writing and the writing next to the line of reasoning recorded on the card also advertises and registers its own references (and indebtedness) with regard to the rest of the complex contents, the user reads along not with his own memory, but rather with his own gradually shifted horizon. It is not only the difference in understanding over time that is surprising. Particularly surprising are the references presented whose paths were even less branched when the thought was entered than at the moment of renewed access. The system of notation continues to develop undetected. Thus, the slip box no longer provides simply the preparatory work for a text to be written, but rather always a kind of pre-form of the text itself. The cross-reference creates from itself, so to speak, the added argumentative value of a slipbox, in that it relentlessly fixes the associations of its reader within a firm association.

Writing and excerpting form the basis of everyday note making. Insofar as such an excerpt always offers a more or less small segment of the source text, it refers to a context that is not collected along with it, but is nevertheless carried over at least as an address. In other words: the excerpt is a pointer that also and always refers to some-thing else. However, an excerpt alone, much less in its referencing function as a proxy, does not make a collection that has a specific productive power. What good is the most meticulous transcript if it cannot be brought into productive relationships with other entries? What good are pages-long excerpts if they do not inscribe themselves into a network of pre-forming cross-references?

It is only with this skill that the slip box moves from the position of a simple filing instrument to the status of an assistant nearly equal to the user, even the position of a proper communicative partner during textual production. For the apparatus does not merely reliably reproduce all that which the user has incrementally invested in it, it recalls the present stretched back to the time of the particular entry. Insofar as it con-nected the material with the previous resources in the entry of excerpts—that is to say, indicated all manner of connections to similar texts, themes, and books—the slip box, as a cross-reference-producing and thus creative guide, delivers numerous connections in their broad branches as simply new, forgotten, or unforeseen lines of thought. Thus, the inconspicuous but consistent cross-reference produces fertile excesses, in that the recombinatory logic of linkage enhances the power of the excerpts through intercon-nected chains of reference.

"It's a shame about your valuable time,"[53] declares the advertising campaign of an office organization firm, which pitches an integrated workstation and administrative environment (not coincidentally) in the journal *The System*, including a desk with in-house mail, files, and a "card index close at hand." Contrary to the defiance of index-theoretical imperatives established thus far, the labor organization and economy under Bielefeld 1951ff. actually obeys the discourse of streamlining in the Weimar period. Luhmann: "The one thing that is really a nuisance is this lack of time."[54] According to the Taylorism movement, a well-organized workplace in particular promises to make daily actions easier and thus allow more time and commitment for the important tasks. "The slip box costs me more time than writing books."[55] Bearing in mind the fact that more information is extracted from than injected into the slip box, since the linkage creates an added value, it is not surprising how the manpower is divided in the Bielefeld 1951ff. system. Most of the read/write head's attention goes to the care and information supply of the system of notes, during which reading is subject to the necessities of efficiency. "H: What happens if you read a book a second, a third time? Do you then take the old card and expand it? L: Sometimes I do it all over again, but that happens relatively rarely. I mean generally, repeated reading happens relatively rarely. H: Really? L: Yes. H: Even the great standard works, for example Krisisschrift, Husserl . . . L: Yes, that . . . I don't read it again, I mean, I could, but I just don't have any time, I have to . . . my reading is always problem-oriented and . . . H: Yes, that is to say, whatever is recorded in this slip box as read and understood, even if it was the greatest error, stays. L: Yes, yes. H: Until someone opens your eyes, so to speak, and says: 'My friend, it simply doesn't say that. You just read that into it.' L: Yes, yes. Then I can note that on the card."[56]

In contrast to the typewritten manuscripts of books, after 1951, the slips in Bielefeld are written by hand, which contributes to a nearly unbridgeable process of exclusion: apart from the necessary graphological barrier for unauthorized readers, which already constitutes an initial hurdle, the togetherness of a similar psychic system and the system of notes derived from it seems highly idiosyncratic, that is, unreadable, because it is incomprehensible for others. The partner does not speak with everyone, but rather only with one. In that third parties remain largely excluded from access and thus from ownership of obvious information, a great value arises from the system that is not merely measured externally. Nevertheless, the system of notes proves to be equally inaccessible to its owner. For its part, the information is protected through the hermeticism of *numerus currens* as well as hidden by the wooden front side of the box. The increasingly complex system conceals its entries successively in the total abundance of material, which can only be exploited by means of a registry. It is the courage of jumping in at a promising point that sets the inquiring reader on the path to a referential journey, a browsing process that opens up new perspectives on the initial question with each new card. The secret, as Georg Simmel put it, or "the concealment of realities through negative or positive means, is one of the greatest accomplishments of humanity; . . . a tremendous amplification of life [is] achieved through the secret, because much of its contents simply cannot appear in full publicity. The secret offers the possibility of a second world, so to speak, apart from the apparent one, and the former is influenced by the latter in the strongest terms."[57]

The open secret of the system of systems theory lies in the complexity of its equipment. Accumulating and creating connections,[58] a recording system led according to simple, consistent rules—duration: thirty years—generates that complexity which is then diminished by the ensemble of various questions in the form of essays and books. The clandestine fosters above all the desire to look behind the opaque drawer fronts, to tear through the cards in the search for truth, which always beckons as an exciting temptation in the form of a drawer to be opened. A slight hesitation, and there is already an irritation, whether the wood will endure another touch and allow us to trace a piece of information needing to be understood. "And yet the hand fumbles once again for the slip box—2 varieties sit inside, ready for notes: DIN A 9 (36.16x52.56) and DIN A 8 (74.33x52.56); and that is also nothing less than pedantry; rather, simply a question of experience: it depends on the temperment, how long the series of keywords is that one needs for the notation of an impression; in any case, a little DIN A 8 note, with shorthand scribbled in tiny cursive on the front and back (hi! The many 'i'-signs!), corresponds to a quarter page,"[59] as Arno Schmidt wraps his own passion in words.

The beauty and elegance of the complex contents of Bielefeld 1951ff., its simple internal organization and the convenient material encoding in wood, paint and cards promise to continually follow each operational success with more searching questions. Luhmann: "In a certain sense, then, the slip box is a reduction for the construction of complexity."[60] Which is why it is hardly a surprise how many others have succumbed to the temptation of the model. "And so Fred picks himself up every evening from his highly complex clerical work, stumbles, still somewhat confused, to his car and drives straight across the city to reduce complexity at a pair of shapely breasts."[61]

Notes

1. Karl Rosenkranz, *Georg Wilhelm Friedrich Hegels Leben*, 2nd unrevised reprographic ed. (Darmstadt: Wissenschaftliche Buchgesellschaft, 1844/1969), 12f. For a historical justification for Hegel's manuscript/typographical idiosyncrasy: "Hegel writes, as is common in Germany, in

Gothic script. For other languages, whether it be Latin, French, English or Italian, he uses the Latin script with Antiqua letters, Greek script for Greek quotes. In a running text, he changes the script when the language changes" (Dierk Schnitger, *Die Papiere und Wasserzeichen der Hegel-Manuskripte. Analytische Untersuchungen*, vol. 4 of *Kataloge der Handschriftenabteilung / Staatsbibliothek Preußischer Kulturbesitz* [Wiesbaden: Otto Harrassowitz, 1995], 21).

2. See Eva Ziesche, *Der handschriftliche Nachlaß Georg Wilhelm Friedrich Hegels und die Hegel-Bestände der Staatsbibliothek zu Berlin Preussischer Kulturbesitz. Teil 1. Katalog*, vol. 4 of *Kataloge der Handschriftenabteilung / Staatsbibliothek Preussischer Kulturbesitz* (Wiesbaden: Otto Harrassowitz Verlag, 1995).

3. Ibid.

4. Niklas Luhmann, *Archimedes und wir. Interviews*, ed. Dirk Baecker und Georg Stanitzek, vol. 143 (Berlin: Merve Verlag, 1987), 149f.

5. Niklas Luhmann, "Kommunikation mit Zettelkästen," in *Ein Erfahrungsbericht*, ed. André Kieserling (Bielefeld: Universität als Milieu, Haux, 1993), 53–61, at 55.

6. Niklas Luhmann, "Lesen lernen," in *Short Cuts* (Zweitausendeins: Frankfurt am Main, 1995 / 2001), 150–156, at 156.

7. Robert Spaemann, "Niklas Luhmanns Herausforderung der Philosophie," in *Paradigm Lost, Über die ethische Reflexion der Moral. Rede anlässlich der Verleihung des Hegel-Preises 1989*, vol. 797 of *suhrkamp taschenbuch wissenschaft*, ed. Niklas Luhmann (Frankfurt am Main: Suhrkamp Verlag, 1990), 49–73, at 62.

8. Niklas Luhmann, *Die Gesellschaft der Gesellschaft*, vol. 1360 of *suhrkamp taschenbuch wissenschaft* (Frankfurt am Main: Suhrkamp Verlag, 1998), 11.

9. See Ernst Robert Curtius, "Goethes Aktenführung," in *Kritische Essays zur europäischen Literatur* (Bern: Francke, 1954), 57–69.

10. "Is the Machine Man's Friend or Man's Enemy?" asks the lead article in the journal *Office Organization* under the headline "Does the Office Machine Devour Souls?" Anonymous, "Frißt die Büromaschine Seelen?" *Büroorganisation* 5, no. 10 (1931): 73, to which there is a (new) objective answer: "The machine need not be an enemy, but also no object of veneration. It has taken over other powers, but created no new ones. The machine is comprehensible, it is no mystical object for the mechanic. Why is it for writers?" (Friedrich Sieburg, "Anbetung von Fahrstühlen," *Die literarische Welt* 2, no. 30 [1926]: 8).

11. Luhmann, "Kommunikation mit Zettelkästen," 53.

12. Ibid.

13. Luhmann, *Archimedes und wir*, 142.

14. Ralf Klassen, "Bezaubernde Jeannie oder Liebe ist nur ein Zeitvertreib," in *Wir Fernsehkinder. Eine Generation ohne Programm*, ed. Walter Wüllenweber (Berlin: Rowohlt Berlin Verlag, 1994), 81–97, at 84.

15. Niklas Luhmann, "Kommunikation mit Zettelkästen. Ein Erfahrungsbericht," in *Öffentliche Meinung und sozialerWandel: Für Elisabeth Noelle-Neumann*, ed. H. M. Keplinger, K. Reumann, and H. Beier (Opladen: Westdeutscher Verlag, 1981), 222–228, at 222. Reprinted in Luhmann, "Kommunikation mit Zettelkästen," 53.

16. Arno Schmidt, "Der Platz, an dem ich schreibe," in *Essays und Aufsätze 2*, vol. 3, 4 of *Bargfelder Ausgabe* (Zurich: Haffmans Verlag, 1995), 28–31, at 28.

17. See Victor Vogt, *Die Kartei. Ihre Anlage und Führung*, vol. 5 of *Orga-Schriften*, 2nd ed. (Berlin: Organisation Verlagsanstalt, 1922), 7f.: "A limp box is quickly worn out, it quickly becomes unsightly and then no longer allows the rapid and secure browsing which is the prime requirement for convenient card index management. . . . With index cards, which are often in daily use for years, the cheapest is always the best."

18. See Deutsches Institut für Normung e.V. (DIN), *Publikation und Dokumentation 2. Erschließung von Dokumenten, Informationsverarbeitung, Reprographie, Bibliotheksverwaltung, Normen*, vol. 154 of *DIN-Taschenbuch*, 2nd ed. (Berlin, Köln: Beuth, 1984), 64f.

19. Wolfgang Hagen, *Warum haben Sie keinen Fernseher, Herr Luhmann? Letzte Gespräche mit Niklas Luhmann* (Berlin: Kulturverlag Kadmos, 2004), 103f.

20. Luhmann, "Kommunikation mit Zettelkästen," 55.

21. Ibid., 55f.

22. Luhmann, *Archimedes und wir*, 145.

23. Luhmann, "Lesen lernen," 156.

24. See Friedrich Kittler, *Die Nacht der Substanz* (Bern: Benteli Verlag, 1989), 15.

25. For Herder, see Nikolaus Wegmann and Matthias Bickenbach, "Herders 'Reisejournal,'" Ein Datenbankreport, *DVJs* 1, no. 3 (1997): 397–420, at 404.

26. Luhmann, *Archimedes und wir*, 144.

27. See Alan M. Turing, *Intelligence Service: Schriften*, ed. Bernhard Dotzler and Friedrich Kittler (Berlin: Brinkmann & Bose, 1987); and for a detailed prehistory see Bernhard J. Dotzler, *Papiermaschinen.Versuch über Communication & Control in Literatur und Technik* (Berlin: LiteraturForschung, Akademie Verlag , Berlin (1996), esp. 7.

28. "If I have nothing else to do, then I write the whole day; in the mornings from 8:30 am until midday, then I briefly go walking with my dog, then I have time again in the afternoon from 2 pm until 4 pm, then it's the dog's turn again. . . . Yes, then I write again in the evenings,

as a rule, until around 11 pm. At 11 pm I mostly lie in bed and read a few more things." Luhmann, *Archimedes und wir*, 145; my emphasis.

29. Ibid.; also Andrew Hodges and Alan Turing, *The Enigma*, vol. 1 of *Computerkultur*, 2nd ed. (Wien, New York: Springer-Verlag, 1994), 115ff.

30. See Vannevar Bush, "As We May Think," *The Atlantic Monthly* 15, no. 176 (1945): 101–108.

31. For one such attempt to expand upon Bielefeld 1951ff. and bring it into electronic form, see synapsen, http://www.verzetteln.de/synapsen.

32. Amid his wandering, he always kept these incunabula of his education. They lie partly in portfolios, partly in cases, on the backs of which a label was glued for orientation. Rosenkranz, *Georg Wilhelm Friedrich Hegels Leben*, 12.

33. Hagen, *Warum haben Sie keinen Fernseher, Herr Luhmann?*, 107.

34. Klassen, "Bezaubernde Jeannie oder Liebe ist nur ein Zeitvertreib," 84.

35. See Niklas Luhmann, *Soziale Systeme. Grundriß einer allgemeinen Theorie*, vol. 666 of *suhrkamp taschenbuch wissenschaft* (Frankfurt am Main: Suhrkamp Verlag, 1984/1994), 346ff.

36. Luhmann, "Kommunikation mit Zettelkästen," 53.

37. For the fundamentally unproductive maintenance of the difference between man and machine from a systems theoretical perspective, see Peter Fuchs, "Kommunikation mit Computern? Zur Korrektur einer Fragestellung," *Sociologia Internationalis* 29 (1991): 1–30, esp. 8f.

38. For the only ontological assumption of systems theory, see the first sentence of the first chapter of the outline of a general theory, Luhmann, *Soziale Systeme*, 30.

39. Ibid., 240, my emphasis of the term, which (not coincidentally?) refers to the Turing machine.

40. Luhmann, *Die Gesellschaft der Gesellschaft*, 82.

41. See first the foundational chapter in Luhmann, *Soziale Systeme*, 191ff. In addition, on the question of the extent to which communication applies to computers, see Peter Fuchs, "Kommunikation mit Computern? Zur Korrektur einer Fragestellung," *Sociologia Internationalis* 29 (1991): 1–30.

42. Luhmann, *Die Gesellschaft der Gesellschaft*, 82f.

43. Heinrich von Kleist, "Über die allmähliche Verfertigung der Gedanken beim Reden," in *Sämtliche Werke und Briefe. Zweiter Band*, ed. Helmut Sembdner (München: dtv, 1805/2001), 319–324, at 319.

44. Ibid., 320.

45. Immanuel Kant, *Metaphysik der Sitten, Zweiter Teil, II. Ethische Methodenlehre*, 1st Section, §50: "He is the midwife of his thoughts," on the teacher-student relationship.

46. Luhmann, "Kommunikation mit Zettelkästen," 57.

47. Walter Benjamin, *Einbahnstraße*, in *Gesammelte Schriften*, vol. 4 (Frankfurt am Main: Suhrkamp Verlag, 1928/1981), 98–140, at 103.

48. Kleist, "Über die allmähliche Verfertigung der Gedanken beim Reden," 323.

49. Johann Wolfgang von Goethe, *The Autobiography of Johann Wolfgang von Goethe*, vol. 2, trans. John Oxenford (Chicago: University of Chicago Press, 1974), 311f.

50. Kittler, *Die Nacht der Substanz*, 16. For a more strictly computer-archeological reading, see also Kittler "Memories Are Made of You," in *Schrift, Medien, Kognition. Über die Exteriorität des Geistes*, vol. 19 of *Probleme der Semiotik*, ed. Peter Koch and Sybille Krämer (Tübingen: Stauffenburg Verlag, 1997), 187–203, at 195–197.

51. Luhmann, "Kommunikation mit Zettelkästen," 58.

52. Ibid., 59f.

53. Advertisement for the firm Hinz, in *Das System. Zeitschrift für Organisation*, no. 1 (January 1928).

54. Luhmann, *Archimedes und wir*, 139.

55. Ibid., 143.

56. Hagen, *Warum haben Sie keinen Fernseher, Herr Luhmann?*, 106f. For literary historical evidence of the extent to which reading isn't reading at all with Jean Paul, see Georg Stanitzek, "'0/1,' 'einmal/zweimal'—der Kanon in der Kommunikation," in *Technopathologien*, vol. 7 of *Materialität der Zeichen, Reihe A*, ed. Bernhard J. Dotzler (München: Wilhelm Fink Verlag, 1992), 111–134.

57. Georg Simmel, *Soziologie. Untersuchungen über die Formen der Vergesellschaftung*, vol. 11 of *Georg Simmel Gesamtausgabe*, ed. Hg. von Otthein Rammstedt, 2nd ed. (Frankfurt am Main: Suhrkamp Verlag, 1908/1995), 405f.

58. Every reference is an interface.

59. Schmidt, "Der Platz, an dem ich schreibe," 32.

60. Luhmann, *Archimedes und wir*, 149.

61. Klassen, "Bezaubernde Jeannie oder Liebe ist nur ein Zeitvertreib," 84.

7 Dataveillance and Countervailance

Rita Raley

It's what we call a massive data-base tally. Gladney, J.A.K. I punch in the name, the substance, the exposure time and then I tap into your computer history. Your genetics, your personals, your medicals, your psychologicals, your police-and hospitals. It comes back pulsing stars. This doesn't mean anything is going to happen to you as such, at least not today or tomorrow. It just means you are the sum total of your data. No man escapes that.

—Don DeLillo, *White Noise*

What is most unfortunate about this development is that the data body not only claims to have ontological privilege, but actually has it. What your data body says about you is more real than what you say about yourself. The data body is the body by which you are judged in society, and the body which dictates your status in the world. What we are witnessing at this point in time is the triumph of representation over being.

—Critical Art Ensemble

The Data Bubble

As I set about the process of wiping my machine of all cookies a few summers ago in preparation for the cloning of my hard drive, I was somewhat naively surprised to learn about so-termed Flash cookies, or LSOs (local shared objects). Internet privacy has always been a concern: I have long made a point of systematically deleting cookies along with my cache and search history, researching the plug-ins and extensions best able to anonymize my browsing, and using search engines that do not record IP addresses, particularly those that work against search leakage.[1] I have also made a point of providing false personal information and developing a suite of pseudonymous identities (user names, avatars, anonymous email addresses), the purpose of which has been to convince myself that I am able to maintain some aspect of control over my own data. My error was thinking within the architecture of the browser window: LSOs are, as the name suggests, local cookies stored outside of the browser, in my case at rraley/Library/ Preferences/Macromedia/Flash Player/#SharedObjects, and thus not deletable from a browser toolbar.[2] In basic terms, LSOs are tracking devices within the Flash player that override the user's security preferences and are set without her knowledge and consent. There are applications such as Flush and BetterPrivacy that will ostensibly manage and clean out LSOs, but their most pernicious aspect is their capacity to "respawn" tracking cookies with data stored in Flash; that is, Flash Local Storage is used to back up the HTML cookies for the explicit purpose of restoring them within seconds after they are deleted. These zombie cookies—and this is certainly their effect, as manual deletion and even the aforementioned tools are essentially futile—are made possible by what Adobe Systems insists is a "misuse" of Local Storage, though it is worth noting that the privacy settings panel on Adobe's site is notoriously difficult to read, appearing as a demo rather than as an actual window.[3] They have not then been invisible to me alone, though the larger issue of data collection continues to receive more public attention in the wake of investigative reports such as the *Wall Street Journal's* "What They Know" series.[4]

The immediate purpose of LSOs, along with traditional and third-party cookies, is online behavioral advertising, the economic potential of which is no doubt clear: consider the speculative value of the uniquely numbered cookie assigned to each machine, one that collates ostensibly nonpersonal behavioral information in order to produce a closely approximate demographic portrait including age, gender, location, educational level, income, consumption habits (purchasing and reading), sexual preference, and health issues.[5] The "audience management experts" of Demdex, Inc., for example, transform the profile of a common user into one of a unique individual by combining the ID code from a single machine, one that holds a summary record of browsing and search history, with offline data including census information, real estate records, and car registration.[6] As John Battelle puts it, this information is producing "a massive clickstream database of desires, needs, wants, and preferences that can be discovered, subpoenaed, archived, tracked, and exploited for all sorts of ends."[7] Online behavioral advertising produces a dynamic, flexible, and perfectly customized audience, constituted by the microtargeting of the intents and interests of consumers on a massive scale. In practical terms, if a consumer happens upon but fails to make a purchase from a particular retail site that aligns with her profile, that microtargeting can become retargeting, which means that ads for an item she has viewed will be pushed to other nonretail sites, or to adopt the rhetoric of personalized retargeting companies, she will be found as she browses and driven back to the original site. In its ultimate form, such a targeting system would locate a user in close proximity to a shopping market, assess the whole of her shopping history, compare those purchases with those of other shoppers, and then push coupons based on that correlated search directly to her mobile device. And that vision is precisely what is driving the current data bubble, in which online behavioral advertising is overvalued, data brokers calculate the speculative futures of data (hedging bets on the unknown uses to which it will be put), and new computational systems are designed to manage both these speculations and the data sets themselves.

We are thus in the midst of what is exuberantly called a "Data Renaissance," in which new marketing worlds await exploration and raw material—raw data—awaits extrapolation, circulation, and speculation. Data has been figured as a "gold mine" and as "the new oil of the Internet and the new currency of the digital world," the engine driving our latest speculative bubble.[8] (Around the time of the worldwide financial crash, venture capital began pouring into online tracking.[9]) Data speculation means amassing data so as to produce patterns, as opposed to having an idea for which one needs to collect supporting data. Raw data is the material for informational patterns still to come, its value unknown or uncertain until it is converted into the currency of information. And a robust data exchange, with so-termed data handlers and data brokers, has emerged to perform precisely this work of speculation. An illustrative example is BlueKai, "a marketplace where buyers and sellers trade high-quality targeting data like stocks," more specifically, an auction for the near-instant circulation of user intent data (keyword searches, price searching and product comparison, destination cities from travel sites, activity on loan calculators).[10] If the catalog era depended on a stable indexical link between data and subject, the behavioral data banks of the present need repeatedly to enact that link through database operations that are not incidentally termed "join" and "union." In other words, my data does not need to be stabilized as a composite profile subject to the interpretive work of personality analysis and motivation research; what matters is simply its functionality in a particular context at a particular moment. In 1993 Critical Art Ensemble suggested that we might begin to thwart the then-emergent data systems by contaminating them with corrupt or counterfeit data.[11] However, data can no longer lose "privilege once it is found to be invalid or unreliable," as they suggest, not only because its truth is operational—if it works it is good—but also because its future value cannot now be calculated. That is, it awaits the query that would produce its value. Data cannot "spoil" because it is now speculatively, rather than statistically, calculated.[12]

The name for the disciplinary and control practice of monitoring, aggregating, and sorting data is *dataveillance*, named as such by Roger Clarke, who suggested nearly twenty-five years ago that it was then "technically and economically superior" to the two-way televisual media of George Orwell's fictional universe.[13] It is such because dataveillance operations do not require a centralized system, provided a set of different databases are networked and provided that they share the same means of establishing individual identification, so that a single unit (an individual or number) can be identified consistently across a range of data sets with a primary key. Dataveillance is not new to information technologies and certainly one could construct a genealogy of biopolitical management that would include paper-based techniques such as the U.S. census. Indeed, in an early commentary on the "electronic panopticon," David Lyon suggests that the difference made by information technologies is one of degree not kind, that they simply "make more efficient, more widespread, and simultaneously less visible many processes that already occur."[14] However, one could argue that there have been qualitative as well as quantitative shifts in dataveillance practices in the last decade, or, more precisely, that an intensification of quantitative differences allows for the articulation of qualitative difference. Dataveillance in the present moment is not simply descriptive (monitoring) but also predictive (conjecture) and prescriptive (enactment). To invoke Gilles Deleuze on the emerging structures of continuous control and assessment, "the key thing is that we're at the beginning of something new."[15]

The question then becomes: what are the materially distinct features of the new unified and dynamic dataveillance regime? Large-scale data-aggregating corporations such as Acxiom and ChoicePoint and increasingly sophisticated tracking technologies such as Flash cookies and beacons indicate a shift in scale, while the emergence of data exchanges indicate a shift in the evaluation and "appreciation" of data itself.[16] The linking of databases, corporate actors, and institutions—as is made possible by corporate acquisitions of DoubleClick (Google) and ChoicePoint (the parent company of Lexis-Nexis)—radically changes the scope of a query, as would the realization of a vision of data storage "measured in petabytes."[17] Speculation lurks here in the incalculable, the size of data storage exceeding conventional metrics and simply open to an unknowable future. Thus is it necessarily the case that data markets should be speculative, their units of exchange not even stabilized as such, and driven by techniques of "predictive optimization" that attempt to generate future value.[18]

Data Subjects

The syncing of browser history with personal and application data has successfully and for the most part uncontroversially been situated under the rubric of "enhanced user experience." Apart from the brief bursts of quasi-theatrical collective outrage—we are shocked to hear Google CEO Eric Schmidt remark that "we don't know enough about you. That is the most important aspect of Google's expansion" or to learn of Facebook's creative interpretations of privacy—there seems to have been a general acquiescence to the notion that the distinctions between private and public and personal and non-personal when it comes to data are at best tenuous and that it is practically and economically in our interest to regard them as such.[19] Indeed, even as the *Wall Street Journal* starkly warns its readers to attend to the question of "What They Know," it continues to speculate on the economic growth potential of data mining. The tone and tenor of comments in user forums ranging from Yahoo Answers to Mozilla Support and Comput ing.net is remarkably consistent: there are basic steps one can take to delete cookies, but it seems unnecessary to do so because they do not interfere with everyday computer use; in fact, some of them are functionally necessary and the end result is that one encounters advertisements that may be of interest. In order to receive customized rather than generalized services, one of course has to provide information to corporations and institutions so that they might better support our preferences, profiles, and favorites. After all, this line of thinking holds, do we not want a personalized Internet that adapts to our individual tastes, habits, and preferences? That it is even possible to speak in such general terms about conditioned behavior is evinced by the memes that play with Google's predictive text feature: What does it think I want when I type "cow"? What does it think my friends want? What mark of distinction accrues to me if the first result is "cowboy bebop" as opposed to "cow clicker"? Such information is shared, circulated, and entered into the field of communicative exchange. In this respect, dataveillance takes its place among affect-generating mechanisms such as Facebook: voluntarily surrendering personal information becomes the means by which social relations are established and collective entities supported. Does this, however, necessarily mean that resignation and ironic acceptance of the new data economy are the doxa?

Pointed questions about behavioral targeting will reveal a certain discomfort from a representative segment of the population; for example, 66 percent of a survey population of adult Americans indicated that they did not want personalized advertising, a number that grew to 73–86 percent when participants were told exactly how companies collect data for targeted ad campaigns.[20] In spite of this, however, the general claim can still be upheld: if in response to the proposed National Data Center in the mid-1960s there was a significant pushback from Congress, the mass media, legal scholars, and the public, in the present moment Americans on the whole seem not to mind being mined.[21] It might then at first glance seem to be possible to speak, as does Mark Poster, of our "interpellation" by databases. True interpellation—in his terms "a complicated configuration of unconsciousness, indirection, automation, and absentmindedness"—requires a coercive system, a "superpanopticon," capable of rendering us as both subjects of and subjects to that particular assemblage that David Mitchell, in a fictional context, calls a corpocracy.[22] For Kevin Robins and Frank Webster, this is the essence of "cybernetic capitalism," by which they mean the whole of the socioeconomic control system that is in part dependent on the capacity of state and corporate entities to collect and aggregate personal data to the extent that each individual can be easily monitored, managed, and hence controlled.[23] As my epigraphs indicate, Robins and Webster are far from alone in their concern with our dynamic incorporation within a totalizing technological system of data management.[24] Greg Elmer also explicates the techniques by which consumer profiles are developed and individuals are "*continuously* integrated into a larger information economy and technological apparatus."[25] But for Elmer and Lyon and others, a crucial aspect of this incorporation is our voluntary participation: the composition of consumer profiles in part results from solicitation—whether in the form of a request for feedback or personal data so as to be granted access to a particular service or program—which means we are interpellated as "self-communicating" actors.[26] To be sure, to participate in the project of modernity has arguably always meant that one becomes a calculable subject by voluntarily surrendering data. Note the established meaning of "datum" itself as it is recorded in the *Oxford English Dictionary*: "a thing given or granted; something known or assumed as fact, and made the basis of reasoning or calculation." In the specific context of a sociotechnological milieu organized according to the operational principles of "cybernetic capitalism," however, our acts of participation or self-communication themselves become data, the entirety of our everyday life practices subject to, and constituted by, perpetual calculation. What was speculative at the time of Don DeLillo's *White Noise* (1985)—"you are the sum total of your data"—has in the intervening years become actualized, and neither the legal nor the political infrastructure has kept pace with the technology.[27]

In December 2009, Google announced that search would thereafter be personalized according to fifty-seven signals, among them location, machine and browser information, and prior search history.[28] The company soon assured its users that it was "recognizing your browser, not you," but who or what is meant by "you" in this formulation? In one account, the "you" is our "data double." Kevin D. Haggerty and Richard V. Ericson explain:

> Surveillance technologies do not monitor people *qua* individuals, but instead operate through processes of disassembling and reassembling. People are broken down into a series of discrete informational flows which are stabilized and captured according to pre-established classificatory criteria. They are then transported to centralized locations to be reassembled and combined in ways that serve institutional agendas. Cumulatively, such information constitutes our "data double," our virtual/informational profiles that circulate in various computers and contexts of practical application.[29]

Financial, travel, and governmental databases might be coordinated but our "data doubles" are only temporarily aggregated, our user profiles produced as an effect or consequence of search queries rather than preexisting stable entities that are then subject to search. It is at this point then that the interpellation argument falters because the processes of subjectification at the heart of the "panoptic sort" have been transformed. Along the same lines, Matthew Fuller argues that surveillance is no longer about visual apprehension but is instead a "socio-algorithmic process" that captures and calculates "flecks of identity," the data trails of our everyday actions, such as our browsing history, financial transactions, and our movements as they are recorded by GPS coordinates on our mobile devices and RFID tags in passports and identity cards.[30] The "flecks" concept emerges in some respect from Gilles Deleuze's outline of the emergence of the "dividual" in the context of the control society; if the individuated self was both product and figure of modernity, "dividuals" are rather fragmented and dispersed data bodies. They are, as Tiziana Terranova explains, "what results from the decomposition of individuals into data clouds subject to automated integration and disintegration."[31] Put another way, they are the CDOs (collateralized debt obligations) of the data market, in which bits and pieces of a supposed composite profile, which is itself an operative fiction, are sliced and diced into different tranches, such that a stable referential link to a singular entity becomes lost in a sea of user intent data. The now-orthodox market position is that the value of data does not depend on its connection to an actual person, until expedience requires that a claim be made for the truth of that data. Our data bodies then are repeatedly enacted as a consequence of search procedures. Data is in this respect performative: the composition of flecks and bits of data into a profile of a terror suspect, the re-grounding of abstract data in the targeting of an actual life, will have the effect of producing that life, that body, as a terror suspect.

Countervailing Engagements

Jack Gladney, the principal character in *White Noise*, is exposed to an airborne toxin and thereafter subjected to a battery of medical tests. The test results are then aggregated with all of his genetic, civic, and personal information to produce a "massive data-base tally," the source and physical location of which are not identified.[32] Gladney considers the conspiratorial implications: "I wondered what he meant when he said he'd tapped into my history. Where was it located exactly? Some state or federal agency, some insurance company or credit firm or medical clearinghouse?" No mere paranoid fantasy, the idea of a single national data center as a matter of public policy was considered during congressional hearings in 1966, with technocratic efficiency weighed against civil liberties, specifically the right to privacy, and a number of representatives expressing concern about the fact that "the computer neither forgives nor forgets" and is "incapable of making allowances for early errors or indiscretions."[33] As Paul Ohm has proven with careful detail, this exact vision of a data bank that "neither forgives nor forgets" is in theory realizable because of reidentification—the reversal of anonymization techniques with such relative ease that anonymization cannot and should not be considered a means of privacy protection.[34] Perfect anonymity is impossible, but the nightmare scenario (then and now) imagines a womb-to-tomb "record prison" or "database of ruin," a massive "database in the sky" held by Google or elsewhere that contains the material necessary to reduce the entropic uncertainty about individual identities and thus cause demonstrable and legally recognized harm to everyone recorded within it. Google's incorporation of DoubleClick, one of the largest behavioral targeting companies, as well as its partnership with Verizon, would likely be the closest approximation of this single database fantasy, but there is as yet no one entity legally (and technologically) capable of aggregating the entirety of "our" data, which would include not only all governmental and financial records but also our entire search and purchase history, along with our relationship to the social graph. (The value at present is in the aggregating of just a few of these data components.) It is the more general sense that data storage is permanent that leads Viktor Mayer-Schönberger to claim that we have been produced as Borgesian figures, like Funes, who have lost the capacity to forget and thereby lost the capacity to structure a temporal narrative.[35] More concretely, the consequence of total storage is that the much-heralded second act of American lives—the mythology of reinvention—cannot be possible if all of the data from the first act is easily accessible.

Data storage of this scale, potentially measured in petabytes, would necessarily require sophisticated algorithmic querying in order to detect informational patterns. For David Gelernter, this type of data management would require "topsight," a top-down perspective achieved through software modeling and the creation of microcosmic "mirror worlds," in which raw data filters in from the bottom and the whole comes into focus through statistical modeling and rule and pattern extraction.[36] The promise of topsight, in Gelernter's terms, is a progression from *annales* to *annalistes*, from data collection that would satisfy a "neo-Victorian curatorial" drive to data analysis that calculates prediction scenarios and manages risk.[37] What would be the locus of suspicion and paranoid fantasy (Poster calls it "database anxiety") if not such an intricate and operationally efficient system, the aggregating capacity of which easily ups the ante on Thomas Pynchon's paranoid realization that "*everything is connected*"?[38]

Happily, sheer impracticality means that data systems can never function as perfectly as our dystopian imaginations might suspect. The errors inherent within a catalog mailing list, one of the more basic data sets, indicates how unstable that data can be: any given population is a massive moving target, all the more so considering the inevitable introduction of false information, and the scale of the sample size—in the TIA topsight scenario, for example, every human entity within the U.S. borders—means that it truly would require the storage of petabytes of data in order to produce accurate calculations. Even if one were to accept the fiction of the universal database managed by a single authority, the fundamental problem of meaningfully, and predictably, parsing that archive remains. Everything might be collected and connected, but that does not necessarily mean that everything can be known. Google may come to possess the sum total of my personal data and all of the history contained within my UID, but it cannot obtain the programmatic perspective necessary to predict exactly what I will buy or what I will read.

Still, as my Firefox add-on, Collusion, reminds me, data collection companies are continually tracking my browsing behavior in spite of my efforts to thwart them, a cogent reminder that targeting is not impractical at the level of the individual. When considered in these terms, it is difficult to dismiss escape, whether in the form of disappearance or disconnectivity, as merely a counterfantasy.[39] Critical Art Ensemble's injunction is to the point: "Avoid using any technology that records data facts unless it is essential."[40] Howard Rheingold and Eric Kluitenberg make a comparable case for "selective connectivity": techniques by which we can "choose to extract ourselves from the electronic control grid from time to time and place to place."[41] Similarly, for Mayer-Schönberger, the solution lies in the adoption of a certain care in the management of one's online interactions, practices of selective disclosure and revelation in order to limit "uncontrollable information flows through individual choice."[42] If we are able to opt out of a single company's personalized retargeting scheme, that is, should we not also be able to opt out of all advertising databases or indeed out of the whole system of "cybernetic capitalism" itself? But it is arguably the case that exit in the form of forgetting or genuine anonymity is no longer possible, that disappearance itself has disappeared. Confronted with this argument we might instead imagine a systems overload, "an information blizzard—a whiteout," because silence can be attained with an increased pitch of white noise.[43] "Anonymity systems function best in a crowd" and therefore overflowing the system, feeding it false information, generating more "flecks of identity" than it can handle, might be the closest approximation of disappearance it is possible to achieve.[44] A creative example of precisely this is Daniel Howe and Helen Nissenbaum's TrackMeNot, a browser extension that works to block the capacity of third parties to identify users based on their search history by periodically creating bursts of search activity and thus hiding real searches within a batch of ghost queries. As the creators explain: "To level the playing field, we have sought to create a mechanism that places some degree of control back in the hands of users and, at every point in the design where this has been feasible, we have sought to do so."[45] Counterpropositions such as these, however, shift the burden of governance from institutions to the mythic entity of the individual rational actor and either argue for or presume a certain technological literacy from the outset.[46] They also imply that data is somehow neutral and that it is only the uses of data that are either repressive or emancipatory.

The critical minefield one must negotiate here is structured by two tried-and-true narratives: one outlining systems of control and the other positioning us as well-informed citizens who can manage (indeed "give") our data and perhaps even turn dataveillance techniques to our own advantage. The version of this binary particular to the Internet pits monopolistic corporations seeking jurisdiction over information architecture and communication flows against those fighting to maintain open, distributed P2P networks (Google is the complicating exception in that it is a single entity whose power derives from the management, support, and ownership of those very distributed networks). If considered in narrowly exclusive terms, each narrative risks a certain blindness: either an overinvestment in the valorization of the agency of the user who hacks the system or an overinvestment in the articulation of the protocols of a given system as inescapably binding, such that it would require naively idealistic faith if not false consciousness to believe in the efficacy and value of resistant and participatory practices. But it remains the case that constellations of control are imbricated with constellations of expressive resistance, whether in the form of tactical intervention, asymmetric infowar, or civic engagement. For every system of disciplinary power, as Anthony Giddens puts it, there is a "countervailing" response from those in precarious, subordinate, or marginal positions, which is to say that dataveillance and countervailance must be seen as inextricably connected.[47] The practices that might be situated under the rubric of countervailance do not endeavor to realize an abstracted philosophy of resistance and human rights. They are often cognizant of such rights, particularly when a governmental program like Poindexter's TIA is articulated within the field of tactical activity as a critical object. But their actions are more often about action itself in relation to a regime that would limit us to efforts to stay on the right side of the data that defines us. Moreover, the expressive aspects of countervailance as I will outline them here serve as an important counter to the technocratic consumer rights initiatives that frame the debate in terms of property—those "MyData" initiatives that seek only to transfer ownership of data to the individual and to develop personal data banks for everyday functionality and monetization.[48]

There are a number of practices that have the potential for disruptive innovation vis-à-vis the new regime of dataveillance. For example, Gary Marx outlines a range of behavioral techniques and legal, economic, and technological exploits ranging from refusal to masking that work toward "neutralizing and resisting the new surveillance" system; neutralization, as he puts it, is a "dynamic adversarial social dance involving strategic moves and counter-moves and should be studied as a conflict interaction process."[49] With respect to consumer (re)targeting and behavioral profiling, a common counter move is the design and programming of anonymizers, encrypters, distributed networks, and ad and cookie blockers. Though many such enterprising programmers may work for large IT corporations, their software can usually be tagged as independent, alternative, open, and almost always free. Just as Internet data mining is dependent on software design, then, so, too, is the blocking or thwarting of that mining. So, to block beacons and zombie cookies and maintain the smallest measure of privacy while reading articles in the *Guardian* online, one can choose from a suite of effective Firefox add-ons including TACO and Beef TACO (targeted advertising cookie opt-out); BetterPrivacy; Ghostery; CookieSafe; and CookieCuller. As Panopticlick, the Electronic Frontier Foundation's browser-fingerprinting algorithm, reveals, however, privacy tools such as spoofers and plugins paradoxically make the browser more distinct and thus facilitate device fingerprinting.[50] Panopticlick further reminds us of the difficulty of demarcating an absolute difference between the means of tracking and the means of circumventing that tracking; another case in point would be browsers in which the facility for private browsing is built into the browser itself.

To understand the significance of software design both to mine and to obstruct, one has only to consider the role that computer models have played in what Andrew Leyshon and Nigel Thrift describe as "the capitalization of almost everything," which is to say in the creation of the explosive development of financial capitalism that led up to the recent global financial crash. In short, "new forms of expertise, fuelled by computing power and software" have been necessarily constitutive.[51] For example, the consolidation and centralization particular to "Shared Services" would not have been possible without the development of a single operating system to aggregate a range of different activities and income streams into a single entity. "As in the case of ground rent, what made the mining of these new seams of financial value [subprime lending] apparently possible is the development of computer software that enables individuals to be assessed, sorted and aggregated along dimensions of risk and reward."[52] Software design did not simply enable the creation of new financial instruments; software design was the essential condition of possibility for these new financial instruments. So, too, the scale and complexity of the data structures at issue—"petabytes"—is such that they cannot be processed by human intelligence alone but rather require machine intelligence in the form of database management systems and algorithms that structure data collection.

Combating or otherwise responding to a control system dependent on computing power requires the design of a counter-system, a rather modest example of which is Diaspora, an open-source, privacy-aware, distributed do-it-yourself social network that eliminates the hub of a social media conglomerate in favor of a peer-to-peer network in which each individual is a node.[53] Without a hub or central server, data encrypted with GNU Privacy Guard is sent directly to one's friends rather than stored and hence mined. True peer-to-peer communication—that is, that which is not routed through a central hub—would need to move to a network such as Diaspora because the controlled application programming interface (API) of social networks such as Facebook means that hacking a hub-based network in order to convert it to peer-to-peer is difficult if not impossible. Regardless, we could point to numerous examples of Facebook users harnessing the peer-to-peer over central hub to mobilize street-based protests, in essence modifying a digitally centralized network so that it functions as peer-to-peer.

Another common countervailing response has been the appropriation of the technological tools of surveillance—whether that be "reciprocal transparency" (watching the watchers) or lateral surveillance, the myriad ways in which people keep track of each other with social networking platforms, cameras, and GPS-enabled mobile devices.[54] Indeed, in the context of social media, lateral surveillance has been considered as a sharing practice involving mutuality and reciprocity rather than a one-way flow of information.[55] So, too, self-directed profiling ("my preferences") means articulating one's own value as a consumer, traveler, citizen, and friend. While dataveillance functions as an instrument of biopolitical control, in other words, it also enables civic participation, at least insofar as one regards as significant the effects of private citizens performing both their own background checks with Google and Facebook and their own market research through user ratings and sites such as Yelp. "Folksonomies," user-created systems for establishing value (via tagging, bookmarking, and rating) similarly function as a means of making community. From Amazon to Digg, there is a vast network to which we can turn to assess our value as producers (of comments, reviews, commodities) and consumers (as trusted users and buyers), one whose seemingly inconsequential rewards (stars, levels) mask a deep sense of community. In this respect, making data public is also making a commons. Apart from functioning as a rival form of expertise, then, one effect of these countervailing tools and techniques has been to re-embed dataveillance within social relations. Perhaps the best example of this is Eyebrowse, a protosocial network based on the self-reporting of one's browsing activities (figure 7.1). A Firefox plugin, Eyebrowse visualizes a user's web browsing history along with that of her friends, thus making visible the data available to Google and any number of third parties, now and in the future.[56]

Mimetically reproducing data collection practices increases technological literacy with respect to both individual everyday practice and systemic logics. Exploiting vulnerabilities makes those vulnerabilities known. Evercookie is perfectly illustrative. The virtuosic work of an elite hacker, evercookie is as it sounds, a tracking device that cannot be destroyed. Designed as a "litmus test," with the tag line "never forget," evercookie provides incontrovertible proof of our relative inability to control the storage of cookies on our computers, particularly in the scripting environment of HTML 5.[57] A more

Figure 7.1   *Eyebrowse*, created by Brennan Moore, Max Van Kleek, and David Karger (MIT CSAIL).

ordinary example is the Firefox extension Firesheep, which allows users to capture the unencrypted login cookies of others on the shared Wi-Fi network, thereby substantiating the need for HTTPS. The hope is that participatory and educative tracking tools such as these produce a more-informed public and blur the lines between a data class that does not understand at a basic level how cookies function and a class of power users savvy enough to exploit the resources at their disposal in the interests of constituting their own data bodies. What becomes apparent after several hours of hands-on work tinkering in search of the perfect combination of antitracking tools, however, is that expert knowledge quickly becomes the aspirational goal, with legal and technological complaints about data mining mollified by the temporary satisfaction of having joined the elite data class. Nonetheless, an embodied experience of dataveillance tools and techniques alerts the public to its role as a stakeholder for, Alberto Melucci notes, "as mere consumers of information, people are excluded from the discussion on the logic that organizes this flow of information; they are there to only receive it and have no access to the power that shapes reality through the controlled ebb and flow of information."[58] A tool such as Eyebrowse certainly gives its users access to data collection processes, though it might well introduce the question of the extent to which we are being asked to immerse ourselves in the dataveillance regime to the point of identification in order to achieve any sort of agential position. Because inhabitation prompts recognition, however, a fully immersive, participatory, and identificatory practice can still function as a means of using a control apparatus against itself.

Mirror Worlds

Artists who appropriate dataveillance techniques and tools as a medium for creative production inform, enlighten, and help us to imagine otherwise by refusing the fantasy of exodus, a withdrawal from a given political, economic, or cultural system predicated on the notion that there is a neutral external vantage point from which one can begin the work of critical assessment.[59] In a very general sense we might term such work immanent critique: art-activism operating within a given structure and inhabiting a particular perspectival frame, whether that be bioartists' hands-on work in the laboratory or hacktivist interventions within networked systems. The paradigmatic instance of an art practice that inhabits a particular perspectival frame would be that of the Yes Men, whose counterfeit performances in the name of entities such as the WTO, Halliburton, and Dow Chemical continue to be mistaken for the real. In work such as this, critique is situated in the act of mimesis, which is not a refusal of "corpocracy" but a reflection in a double sense: mirroring and replication, on the one hand, and critical contemplation on the other. A reiterative aesthetic serves to engage a public with a reflective understanding of the operations of power and control. Its creative, productive, and playful aspects open rather than foreclose lines of inquiry; in its eschewing of a singular and reductive negative judgment, it maintains a purchase on continuous critical assessment. A reiterative aesthetic can be radically transformational precisely because it exists in dynamic interplay with its object; it neither claims a stable outside nor fixes upon a synchronic slide of a system that is the inevitable byproduct of topsight.

The work of the Preemptive Media collective—whose practice includes instructional workshops and the re-engineering of mobile technologies—is particularly apposite for a discussion of dataveillance and tactical countervailance. Preemptive Media's object is to exploit consumer electronics for a larger purpose, to foster not only technological literacy, but also critical consciousness and a kind of low-tech amateurism. In one representative series of performances, called *SWIPE* (2002–2005), the collective installed a functioning bar in galleries and exhibition spaces and opened it up for enjoyment and play.[60] Patrons ordering drinks had their drivers licenses scanned and were given individual receipts detailing the data culled both from the 2D barcode and online search. Computer stations in the bars displayed a web-based toolkit with a data calculator to allow participants to determine the market value of their individual data; they also displayed the decoding application used in the installations and a thorough guide to the process of requesting one's data files from the large data warehouses: ChoicePoint, Acxiom, LocatePLUS, and Experian. The purpose was to encourage consumer awareness of Automated Identification Data Capture technologies (AIDC); to give participants the experience of visualizing their own data; and to facilitate a critical conversation about data mining, transparency, and privacy. Swiping suggests purchasing, as if one uses currency to establish or prove currency, reminding us of the extent to which the value, significance, and indeed existence of the individual body are calculated, even proved, by complex systems of accounting—the precise operation of which remains obscure. But *SWIPE* interrupted the one-way flow of information from evidentiary subject to surveillance mechanism, enacting in the process lateral relations among the participants. As the bar setting indicates, *SWIPE* functioned within a social space, its relational aesthetic true to Nicolas Bourriaud's vision of an artistic praxis that struggles against the reifying and commodifying of social relations by creating a space for "alternative forms of sociability."[61] Even as it introduced a certain defamiliarizing shock in individual participants, then, it was unambiguous about the situation of those participants within a broader political and socioeconomic matrix. As the artists noted: "Our hope is to encourage thinking beyond the individual self ('I do not care if a bar database has my name and address and time of visit . . .') toward understanding databases as a discursive, organizational practice and an essential technique of power in today's social field."[62]

Osman Khan's installation *Net Worth* (2004) was similarly dependent on the gallery visitor's swipe, in this case of a credit or ATM card, in order to mine the identificatory information necessary to perform a Google search to determine search rank and thus, "net worth"[63] Drawing on the familiar practice of egosurfing, the tracing of one's own virtual-physical presence and presumed importance online, this installation articulated a shift from the moment of the televisual record—you don't exist unless the entire world sees your image—to the moment of the database record—you don't exist unless you appear on Google. So, too, *Net Worth* invokes the discourse on reputation and trusted users in its equating of the assessment of net presence with the assessment of the value of the individual. More recently, David Kemp asked 100 people to show him the identification, banking, and loyalty cards in their wallet—"anything that connects to a database"[64]—and then for his installation, *Data Collection* (2010), he used each data set to compose an individual "canvas" with photographic representations of the cards on which all of the personal information is visible, with some cards blacked out on request of participant. A small sampling of dataveillance art, these projects are both tactile and rhetorical, dependent on the gift of data in order to open a space for the critical contemplation of that data. They work with—both exploit and capitalize upon—participants' willingness to share data for no immediately tangible or concrete reward, that is, for no apparent return on their affective and participatory investment. What is illuminated by each is the logic of social media and relational aesthetics, which is to give by sharing.

A skeptical viewer might ask whether such data works are in fact supportive of, and thus insufficiently attentive to, their own corporate and governmental information architecture. But this is a variant of the old worry about artists not having sufficient critical distance from the capitalist, technological, scientific, and ideological systems within which they are working. In other words, to suggest that using data-mining techniques to produce art necessarily entails adopting the very logic and optics of the dataveillance society is to rehash the old problem of disinterest. The common assumption is that distance is necessary for critical reflection and that proximity necessarily produces corruption. But the lesson I think we need to learn from tactical media practitioners more broadly is that critique and critical reflection are at their most powerful when they do not adopt a spectatorial position on the (putatively neutral) outside, when they do not merely sketch a surface, but rather penetrate the core of the system itself, intensifying identification so as to produce structural change.[65] Such a practice—such a mode and positioning—goes well beyond Michel de Certeau's notion of "undermining a system from within"; these are not employees wasting time and using the resources of the workplace to turn it against itself.[66] Rather, these art-activists are creating "mirror worlds," replicating the scene of data mining—swiping an identification card—to enable an embedded and embodied perspective of the control network through and within which dataveillance operates, but without the fantasy of exteriority. Instead, the force of the immanent critique envisioned here derives from a near-total inhabitation of the frame, compelling a jarring recognition from the viewer/user and leading to a temporal interval in which she must formulate a response, whether that be rejection or acquiescence. Interventionist art projects such as these work directly against the forces of interpellation with a counterimage of a dataveillance regime that makes that regime perceptible—and if it is perceptible then it becomes possible to work concretely toward political transformation.

The role played by the designers of countervailing tactics, tools, and techniques is akin to that played by the "Keymaker" in *The Matrix Reloaded*: they offer access to a back door, a shortcut key or authenticating token that holds out the promise of allowing us to circumvent the programmed structure of the dataveillance regime.[67] The film is reflexively archetypal. The Oracle instructs Neo, the One, to find the Keymaker, who is being held captive by a master program because of his knowledge of the rules of the system and his ability to open a door leading to The Source. His pre-scripted function is to sacrifice himself to The Resistance project. When Neo opens the door to his prison to find him in the act of making the single key, he announces his function: "I'm the Keymaker. I've been waiting for you." He tells the skeptical ship commanders that he knows of the door and the building level "where no elevator can go, where no stair can reach" because he "must know" and it is "his purpose," the "reason" he is there. And when he is killed by the agents after opening the door to the antechamber, he tells Neo and Morpheus simply that "it was meant to be." In other words, he is programmed only to exploit the weakness in the system, after which he becomes expendable. Read representationally, the Keymaker program is an integral component of the matrix: control systems must necessarily have moles who can reveal the means of puncturing the system so as to satisfy the demand for breaking through (or leveling up)—a demand that is at once narratological and psycho-social. These acts have precise actors ("only The One can open the door"), precise spatiotemporal coordinates ("only during the window can that door be opened"), precise organizational logics ("All must be done as one. If one fails, all fail"), and they can be performed exactly once. Once the door is opened or the threshold crossed, the act cannot be repeated. The flip side of the fantasy of total information awareness, then, is the fantasy of breach.

But the Keymaker does not need narrative structure to legitimate his energies; indeed he dies even before the plot of which he speaks is realized. His role does not exactly duplicate that of countervailing actors—I am not after all advocating sacrifice—but it is emblematic. On the one hand his knowledge is scripted ("I know because I must know") and his circumvention of the system thus simply an exercise in self-regulation. The extra-institutional spaces, here the hallway that is not legible within the matrix, are themselves built into the system and subject to management. On the other hand, however, the wily Keymaker does elude the agents and open the door, which is to say that the act is neither a protocol nor sabotage but both, and self-reflexively so.

So, too, evercookie, the indestructible cookie, is neither purely a tracking technology nor a hack designed to show vulnerabilities and *SWIPE* is neither actual data collection nor a performance of the same but both/and. In other words, dataveillance and countervailance coexist not in dialectical struggle but rather are so fundamentally entangled that the line separating the one from the other is unstable. Positioned as we are within the dataveillance regime, we cannot but employ the tactics of immanent critique, which depends not on an overstatement or overarticulation of totalizing control systems nor on a hyperbolized romance of the exploitation of these systems, but rather depends simply on ordinary action itself.

Acknowledgments

Many thanks to Russell Samolsky, Lisa Gitelman, Rahul Mukherjee, Juliette Cherbuliez, and Francisco J. Ricardo for careful reading and helpful suggestions. Earlier versions of this paper were presented at the American Political Science Association annual conference in Washington, DC; "DIY Citizenship: Critical Making and Social Media," University of Toronto; and Department of Communication Studies, Concordia University. Thanks to Renee Marlin-Bennett, Megan Boler, Matthew Ratto, and Charles Acland for the invitations to present this work and to the audiences for significant feedback.

Notes

1. "Search leakage" is the disclosure of search terms to visited sites; that is, a record of the path followed to land on a particular page. Search engines that allow one to surf anonymously, most of which neither record IP addresses nor use identifying cookies, include Scroogle, Ixquick, DuckDuckGo, and Yauba. Another way to prevent search leakage is to use network routing software like Tor, an "infomediary" that encrypts traffic between the individual user and the Tor network. More simply, encrypted search (HTTPS, or HTTP secure) does not send search terms. The Electronic Privacy Information Center (EPIC) maintains an extensive list of privacy tools for voice, email, instant messaging, and browsing, as does the Center for Democracy and Technology. See http://epic.org/privacy/tools.html and https://www.cdt.org/privacy/guide/basic/tips.php (accessed February 7, 2011).

2. A 2009 article in *Wired*, admittedly usually a bit delayed both with its techno-boosterism and techno-paranoia, suggests that LSOs have for the most part escaped general notice, a point made in a number of related articles then and since. See Ryan Singel, "You Deleted Your Cookies? Think Again," *Wired* (August 10, 2009), http://www.wired.com/epicenter/2009/08/you-deleted-your-cookies-think-again (accessed August 10, 2009). By the time this chapter makes it into print, it, too, will likely seem a bit belated, particularly as HTML 5 comes into widespread

use, but it can be read as a snapshot account of dataveillance practices and the tactics, techniques, and technologies deployed to negotiate them in the era of big data, a battle that will almost certainly persist for the foreseeable future.

3. Adobe Statement for the Privacy Privacy Roundtables Project filed with the Federal Trade Commission (January 27, 2010), http://www.ftc.gov/os/comments/privacyroundtable/ 544506-00085.pdf (accessed July 25, 2010). Clearspring Technologies, one of the larger content-sharing companies, and the developer of the AddThis platform, discloses its use of Flash cookies in its privacy policy for AddThis, but not on the privacy policy for the company itself. See http://www.addthis.com/privacy (accessed November 14, 2010).

4. At the time of this writing, the "What They Know" section of WSJ.com continues to be regularly updated. See http://online.wsj.com/public/page/what-they-know-digital-privacy.html (accessed February 7, 2011).

5. The concepts of "personal" and "nonpersonal" are, as one would expect, somewhat mutable in the context of dataveillance. The single cookie assigned to each machine is not automatically attached to an individual identity so, while sexual preference might in certain legal statutes be defined as "personal," in the context of information security it would be considered nonpersonal. Personally identifiable information (PII), on the other hand, includes social security numbers, genetic information, biometric data, date of birth, and in some cases vehicle registration numbers, bank numbers, and IP addresses, although the increasingly widespread use of proxies makes the last more complicated. Much of the data-privacy legislation to date restricts the use of PII and presumes the safety of anonymization.

6. The Adobe AudienceManager platform, which is based on Demdex, invites companies to create a data bank based on both their own ad campaigns and data acquired from third parties. See http://www.demdex.com (accessed February 10, 2011).

7. John Battelle, *The Search: How Google and Its Rivals Rewrote the Rules of Business and Transformed Our Culture* (New York: Portfolio, 2005), 6. The structural logic behind online behavioral advertising would be the "panoptic sort," Oscar Gandy's descriptive formulation for the system that "operates to increase the precision with which individuals are classified according to their perceived value in the marketplace and their susceptibility to particular appeals." Oscar H. Gandy, *The Panoptic Sort: A Political Economy of Personal Information* (Boulder, CO: Westview Press, 1993), 2.

8. Julia Angwin, "The Web's New Gold Mine: Your Secrets," *Wall Street Journal* (July 30, 2010). Meglena Kuneva, cited in Marc Davis, keynote presentation, Privacy, Identity, Innovation annual conference (Seattle, 2010), http://vimeo.com/14401407 (accessed November 12, 2010).

9. Scott Thurm, "Online Trackers Rake In Funding," *Wall Street Journal* (February 25, 2011), http://online.wsj.com/article/SB10001424052748704657704576150191661959856.html (accessed November 12, 2010).

10. See http://www.bluekai.com (accessed November 12, 2010).

11. Critical Art Ensemble, *The Electronic Disturbance* (New York: Autonomedia, 1993), 63.

12. Ibid., 140.

13. Roger Clarke, "Information Technology and Dataveillance," *Communications of the ACM* 31, no. 5 (May 1988): 499, http://www.rogerclarke.com/DV/CACM88.html (accessed November 12, 2010). In this chapter I focus specifically on dataveillance in the sense of data mining (capture and aggregation), as opposed to the whole suite of techniques and technologies of a contemporary electronic surveillance regime, ranging from CCTV to biometrics, though they are by no means unrelated.

14. David Lyon, *The Electronic Eye: The Rise of Surveillance Society* (Minneapolis: University of Minnesota Press, 1994), 40.

15. Gilles Deleuze, "Postscript on Control Societies," *Negotiations, 1972–1990*, trans. Martin Joughin (New York: Columbia University Press, 1995), 182.

16. Beacons such as web bugs and pixels track user keyboard and mouse activity on a given webpage.

17. Cited in Elliott Borin, "Feds Open 'Total' Tech Spy System," *Wired* (August 7, 2002), http://www.wired.com/politics/law/news/2002/08/54342 (accessed August 25, 2010).

18. One company, [x+1], has named its product the Predictive Optimization Engine (POE ™). See http://www.xplusone.com/glossary (accessed November 10, 2010).

19. Quoted in Ira Rubinstein, Ronald D. Lee, and Paul M. Schwartz, "Data Mining and Internet Profiling: Emerging Regulatory and Technological Approaches," *University of Chicago Law Review* 75 (2008): 273.

20. We can say, then, that the privacy crisis produced by the new practices of data collection is to a certain extent hidden in plain sight and recognizable only in moments of elucidation. Joseph Turow, Jennifer King, Chris Jay Hoofnagle, Amy Bleakley, and Michael Hennessy, "Americans Reject Tailored Advertising and Three Activities That Enable It" (September 29, 2009), http://papers.ssrn.com/sol3/papers.cfm?abstract_id=1478214 (accessed November 10, 2010).

21. Arthur Miller provides a full account of the proposal and its reception in his seminal text, *The Assault on Privacy: Computers, Data Banks, and Dossiers*, which was itself positioned as a response to the idea. As he notes, the public debate managed to avoid "the fundamental policy issue of how to curtail the government's increasing penchant for information collection" and a defeat of the proposal for the national data network simply meant that each governmental agency developed its own: *The Assault on Privacy* (Ann Arbor: The University of Michigan Press, 1971), 59. In a quite general sense, public reaction needs to be situated in the particular historical,

sociocultural, and juridical contexts of a nation-state. Consider, by contrast, the German government's successful campaign against Google's Street View feature. For an early report on the issue, see Kevin O'Brien, "Google Data Admission Angers European Officials," *New York Times* (May 15, 2010), http://www.nytimes.com/2010/05/16/technology/16google.html (accessed May 15, 2010).

22. Mark Poster, "Databases as Discourse; or, Electronic Interpellations," *Computers, Surveillance, and Privacy*, ed. David Lyon and Elia Zureik (Minneapolis: University of Minnesota Press, 1996), 187. The Sonmi chapters of Mitchell's *Cloud Atlas* are set in the dystopian corpocratic state called Nea So Copros.

23. Kevin Robins and Frank Webster, "Cybernetic Capitalism: Information, Technology, Everyday Life," *The Political Economy of Information*, ed. Vincent Mosco and Janet Wasko (Madison: University of Wisconsin Press, 1988), 44–75.

24. For a thorough account of pattern-based searches by government and corporations and the techniques one can use to mask online activity; a detailed overview of the public outcry over TIA, its subsequent de-funding, and the continuation of the same data-mining exercises under the classified intelligence budget; and, finally, a detailed legal review that makes the case for transparency and new identity technologies with privacy protections, see Rubinstein, Lee, and Schwartz, "Data Mining and Internet Profiling." It is important to note, however, that these debates are premised on a notion of privacy with a particular history and cultural specificity.

25. Greg Elmer, *Profiling Machines: Mapping the Personal Information Economy* (Cambridge, MA: MIT Press, 2004), 17.

26. Lyon, *The Electronic Eye*, 52.

27. There are substantive juridico-political questions that need to be addressed as the legal infrastructure develops: What is the legal status of our financial records, unique ID codes, and biometric data? How or to what extent will individual data be monetized? Can individual browsing be considered labor? If so, would not the unique ID code that records sites visited be considered a product of that labor and thus private property? Does the person from whom data originated have claims over it once it enters into circulation on the "data exchange"? Will data follow the model of genetic materials, with data becoming the intellectual property of a data broker who had altered it in some fashion? Proposed policy solutions thus far include improved securitization, transparency and informed consent, expiration dates and storage limits, and the regulation of data centers.

28. On the era of personalization, see Eli Pariser, *The Filter Bubble: What the Internet Is Hiding from You* (New York: Penguin Books, 2011).

29. Kevin D. Haggerty and Richard V. Ericson, *The New Politics of Surveillance and Visibility* (Toronto: University of Toronto Press, 2006), 4.

30. Matthew Fuller, *Media Ecologies: Materialist Energies in Art and Technoculture* (Cambridge, MA: MIT Press, 2005), 149.

31. Tiziana Terranova, *Network Culture: Politics for the Information Age* (London: Pluto Press, 2004), 34.

32. Don DeLillo, *White Noise* (New York: Viking, 1985), 140.

33. Hearings on the Computer and Invasion of Privacy before a Subcommittee of the House Committee on Government Operations, 89th Congress, 2nd Session (Washington, DC: U.S. Government Printing Office, 1966) 3, 12.

34. Paul Ohm, "Broken Promises of Privacy: Responding to the Surprising Failure of Anonymization," University of Colorado Law Legal Studies Research Paper No. 09-12 (August 13, 2009). http://ssrn.com/abstract=1450006 (accessed August 25, 2010).

35. The reference here is to the Jorge Luis Borges story, "Funes the Memorious." Viktor Mayer-Schönberger makes a case for putting an expiration date on information, which would mean customizing each data element so that it remains accessible for a limited period of time—that is, for flexible implementation rather than a general legal code. See *Delete: The Virtue of Forgetting in the Digital Age* (Princeton, NJ: Princeton University Press, 2009).

36. David Gelernter, *Mirror Worlds: Or, the Day Software Puts the Universe in a Shoebox . . . How It Will Happen and What It Will Mean* (New York: Oxford University Press, 1992). John Poindexter's plans for the Total Information Awareness Program (TIA) drew on Gelernter's paradigm, endeavoring to use the principle of topsight to establish a terror network that could ostensibly be seen and disciplined, though not eliminated because of its regenerative ends.

37. Ibid., 112.

38. Thomas Pynchon, *Gravity's Rainbow* (New York: Penguin Books, [1973] 1995), 703.

39. See Irving Goh, "Prolegomenon to a Right to Disappear," *Cultural Politics* 2, no. 1 (March 2006): 97–114.

40. Critical Art Ensemble, *Electronic Disturbance*, 135.

41. Howard Rheingold and Eric Kluitenberg, "Mindful Disconnection: Counterpowering the Panopticon from the Inside," *OPEN 11 Hybrid Space* (Amsterdam: NAi Publishers, 2007), 32.

42. Mayer-Schönberger, *Delete*, 129.

43. Critical Art Ensemble, *Electronic Disturbance*, 132.

44. Rubinstein, Lee, and Schwartz, "Data Mining and Internet Profiling," 277. It is not uncommon to hear this argument made with respect to social media; in other words, if everyone's intimate details are available, we are essentially hidden in plain sight.

45. TrackMeNot FAQ, http://cs.nyu.edu/trackmenot/faq.html (accessed August 25, 2010).

46. The detailed counsel about risk management online offered by the Electronic Frontier Foundation's Surveillance-Self Defense Project is paradigmatic. See https://ssd.eff.org (accessed October 14, 2011).

47. Anthony Giddens, *The Nation-State and Violence* (Berkeley: University of California Press, 1985), 186.

48. See Richard H. Thaler, "Show Us the Data. (It's Ours, After All)," *New York Times* (April 23, 2011), http://www.nytimes.com/2011/04/24/business/24view.html (accessed April 23, 2011) and "Better Choices, Better Deals," U.K. Cabinet Office (April 13, 2011), http://www.cabinetoffice.gov.uk/resource-library/better-choices-better-deals (accessed July 14, 2011).

49. Gary Marx, "A Tack in the Shoe: Neutralizing and Resisting the New Surveillance," *Journal of Social Issues* 59, no. 2 (May 2003): 369–390.

50. The odds of someone in my time zone using the same browser, operating system, font set, privacy tools, and precise microversions of plugins (Java 1.6.0_17) are remarkably low. Cookies leave crumbs, however dispersed and persistent, while fingerprinting does not, which means that browser tagging essentially goes undetected. Though fingerprints can be associated with search terms, the economic and political utility of fingerprinting, apart from authentication, is not yet entirely clear, which again suggests that what is at stake are potential uses and abuses.

51. Andrew Leyshon and Nigel Thrift, "The Capitalization of Almost Everything: The Future of Finance and Capitalism," *Theory, Culture & Society* 24, no. 7–8 (2007): 101.

52. Ibid., 108.

53. Developed by four Columbia University undergraduates in response to a lecture on Internet privacy, Diaspora is as of this writing still in alpha version. See http://www.joindiaspora.com and Jim Dwyer, "Four Nerds and a Cry to Arms Against Facebook," *New York Times* (May 11, 2010), http://www.nytimes.com/2010/05/12/nyregion/12about.html (accessed May 11 2010).

54. See David Brin, *The Transparent Society: Will Technology Force Us to Choose Between Privacy and Freedom?* (New York: Basic Books, 1999); and Mark Andrejevic, "The Work of Watching One Another: Lateral Surveillance, Risk, and Governance," *Surveillance & Society* 2, no. 4 (2005): 479–497.

55. Anders Albrechtslund, "Online Social Networking as Participatory Surveillance," *First Monday* 13, no. 3 (March 3, 2008), http://firstmonday.org/htbin/cgiwrap/bin/ojs/index.php/fm/article/viewArticle/2142/1949#6 (accessed September 22, 2010).

56. See Marx Van Kleek, Christina Xu, Brennan Moore, and David R. Karger, "Eyebrowse: Real-Time Web Activity Sharing and Visualization," *CHI 2010* (April 10–15, 2010). ACM 978-1-60558–930–5/10/04.

57. Tanzina Vega, "New Web Code Draws Concern Over Privacy Risks," *New York Times* (October 10, 2010), https://www.nytimes.com/2010/10/11/business/media/11privacy .html?hp (accessed October 10, 2010).

58. Alberto Melucci, *Challenging Codes: Collective Action in the Information Age* (Cambridge: Cambridge University Press, 1996), 180.

59. A full account of surveillance art is necessarily outside the scope of this chapter; on this topic, see Thomas Y. Levin, Ursula Frohne, and Peter Weibel's nearly comprehensive *[Ctrl] Space: Rhetorics of Surveillance from Bentham to Big Brother* (Cambridge, MA: MIT Press, 2002). The works I have selected for discussion more narrowly engage the collection of consumer data.

60. *SWIPE*, http://www.preemptivemedia.net/swipe/bar/index.html (accessed February 28, 2012).

61. Nicolas Bourriaud, *Relational Aesthetics* (Paris: Les Presses du réel, 2002), 44.

62. *SWIPE*, http://web.archive.org/web/20060117165314/http://www.we-swipe.us/plain .html#about (accessed February 28, 2012).

63. Osman Khan, *Net Worth*, http://www.todayandtomorrow.net/2005/09/15/net-worth (accessed September 1, 2010).

64. Quoted in Jan Allen, "Sorting Daemons," *Sorting Daemons: Art, Surveillance Regimes and Social Control* (Kingston, Canada: Agnes Etherington Art Centre, 2010), 22.

65. I make this case in *Tactical Media* (Minneapolis: University of Minnesota Press, 2009).

66. Michel de Certeau, *The Practice of Everyday Life* (Berkeley: University of California Press, 1984), 179.

67. *The Matrix Reloaded*, directed by Andy and Lana Wachowski (2003; Burbank, CA: Warner Home Video, 2003), DVD.

8 Data Bite Man: The Work of Sustaining a Long-Term Study

David Ribes and Steven J. Jackson

Introduction

This chapter makes one basic point: the work of producing, preserving, and sharing data reshapes the organizational, technological, and cultural worlds around them. Data are ephemeral creatures that threaten to become corrupted, lost, or meaningless if not properly cared for. Long ago, data managers moved past speaking in narrow technical terminologies, such as "storage" and "transmission," and turned to a more nuanced vocabulary that included "data preservation," "curation," and "sharing." These terms are drawn from the language of library and archival practice; they speak to the arrangement of people and documents that sustains order and meaning within repositories.

In this chapter, we seek to push beyond the commodity fictions of data[1] that too often characterize and limit studies of data sharing. In particular, we tell stories of data production that reveal the complex assemblage of people, places, documents, and technologies that must be held in place to produce scientific data. The vehicle for our discussion is a distinctive case of long-term ecological research: stream chemistry data in the Baltimore region. We follow the practices of scientists and technicians from the field site to shared online repository as the natural world is translated, step by step, from flowing streams to ordered rows of well-described digital data, readily available for use in science.

The data, things, and people we care about here face particular temporal challenges. Data that stretch across years, decades, and ideally centuries are increasingly important within the ecological and climatological sciences that seek to generate "harder" evidence about longitudinal changes in the environment. Data must be comparable across time and sufficiently well described so as to facilitate integration with other data. Our cases demonstrate the staggering amount of work that goes into the production of information for scientific purposes. But even more revealing are the mounting constraints scientists face in seeking to preserve data across time and to collect data that year after year continue to stand in for *the same phenomena*.

Conditions are continuously changing, whether environmental, human, or infrastructural. Sites where samples are collected are transformed over the years, becoming polluted by industrial growth and then purified as emission standards take effect. The academic cycle brings in new teams of graduate students (the laborers of scientific data production) and each such change threatens to tweak the delicate rituals of collection. New sensors promise automation and objectivity while subtly changing baseline readings and the accompanying human routines of collection and upkeep. What results is a complicated ontological choreography,[2] as scientists and technicians work to make data "the same" in a changing ecology of technologies, organizations, field sites, and institutional rearrangements.

In this context, data—long-term, comparable, and interoperable—become a sort of actor, shaping and reshaping the social worlds around them. Demanding and fickle, at the slightest change of condition they threaten to cease being useful for the scientific work they were born to accomplish. To bring us to ecological field sites in Baltimore, we first begin with three accounts of ecologies, scientific objects, and data archives that exemplify the ways phenomena shape the social orders that seek produce, manage, and preserve them. These accounts include (1) corn as a world builder, (2) flies that multiply data, and (3) data that threaten to overheat. We then turn to our detailed empirical analysis of production work for a long-term data stream within the ecosciences.

Corn Thrives in Industrial Ecology

As Michael Pollan describes, corn is an unlikely imperialist.[3] The species that has come to dominate global agriculture struggles to survive in the wild precisely for the reason that we humans find it so useful; with row after row of tightly packed kernels inside a thick protective husk, corn is more likely to rot than thrive in the absence of a creature with opposable thumbs to tear open the husk and individually plant the kernels. Even if an ear of corn somehow manages to lose its husk and fall to the soil, hundreds of seeds will sprout, crowd each other out, and die long before the reproductive cycle is complete. Corn, like more and more species then, has thrown its lot in with humans, adapting to the contemporary social world—and especially to industrial agribusiness— with such success that it has pushed nearly all other staple competitors out of business as a cornerstone of our food supply. Like Britain in its heyday, the sun never sets on the empire of corn.

Pollan offers a compelling picture of the trade routes of the corn empire, documenting the production of a raw ear of corn from a farm in Iowa, and then tracing all the steps it takes as it travels to the typical American consumer. We often think of this end product as the "raw" ears of corn that we purchase at the grocery store and imagine that it is shipped to our stores more or less directly from the farmer. But as Pollan describes, most corn enters our kitchen (and our bodies) through a much more circuitous route—losing its rustic form almost as soon as it is pulled from the ground. In our industrial economy, every portion of the plant is systematically stripped, collated, and processed to produce a standardized set of products, including the now-famous high-fructose corn syrup, cornstarch, MSG, maltodextrin, ethanol, and citric acid. Such derivatives are shipped to ranches and factories across the country where they serve as raw material for new products—constituting the basis of a full quarter of American processed food.

If we have domesticated corn, it has just as surely domesticated us. As Pollan argues, "It takes a certain kind of eater—an industrial eater—to consume these fractions of corn, and we are, or have evolved into, *that* supremely adapted creature: the eater of processed food."[4] And we are not the only species on the planet that has been so domesticated: in one memorable chapter, Pollan details the heroic efforts required to create the now ubiquitous corn-fed American steer, a particular challenge "since the cow is by nature not a corn eater."[5] Other chapters reveal how our financial system has been reconfigured to handle the deluge of industrialized corn, with new technologies like commodity markets and futures trading developed to support the ever-lengthening pathways between farmer and consumer.[6] Even the American farmer, an archetypal figure of autonomy and self-reliance, has been turned into a factory worker at the service of a commodity—corn—most varieties of which can now not even be *eaten* without substantial industrial processing.

As with corn so too with data . . .

Like corn on the cob that arrives to our grocery stores in conditions resembling its state in the field, we often think of raw data as following straight and commonsensical pathways from collection to database. Sometimes this is true (there are still farmer's markets, after all). However, the more common story—especially in today's "big science" projects—is an increasingly Pollan-esque one, with data moving through complex, multi-institutional networks, sharing more similarities with the production of industrial corn than the traditional understandings of field or laboratory science. This is in some ways the *ambition* of contemporary "big science" investments: a more complex, dynamic, and commensurable world in which data really *can* flow freely like corn, leaving new systems, processes, and discoveries in their wake. To do this, we must domesticate data: establishing rituals and routines of collection, creating safe pathways for samples to travel, and setting metadata standards to render them comprehensible by others. And in doing so, data increasingly domesticate us.

Flies Dissatisfied with Information System

As historian of science Robert Kohler describes, the fruit fly *Drosophila* (and its most common lab species, D. Melanogaster) was not born as a laboratory animal per se.[7] Already "cosmopolitan," it has cohabited with us in cities for millennia; it is the fruit fly most likely to appear if you were to put a banana out on your window sill and then wait for the larvae to mature. Breeding ferociously in autumn, it is most plentiful at the beginning of the academic year—just in time for a fresh crop of undergraduate, graduate, and faculty experiments. As raw material for students' projects, it is readily available, cheap, easily maintained, and quickly restocked. Thus, there was always already an elective affinity between the labs of urban bioscience and what would become one of its most common objects. Melanogaster helped create a technology of research on which fly researchers came to depend for their professional livelihood. Once inside the lab, the fruit fly took on a new life of its own and came to drive research at paces never before seen in genetics—eventually demanding novel data management and classification strategies.

Scientists first began to use the fruit fly for genetic research in 1901 at Harvard and since then it has become a dominant species in this new ecosystem: the lab. While capable of sleepily surviving the outdoor winter, Drosophila took to the warmth and security of labs with perennial reproduction. Defining an entirely new criterion of fitness, its productivity in this new ecological niche pushed down the traditional species inhabiting the genetic lab: the rat and mouse, the pea and primrose.

One of the foremost early Drosophila scientists, Thomas Morgan, writing of the relentless reproductive productivity of Melanogaster, enthused: "It is wonderful material. They breed all the year round and give a new generation every sixteen days." As time passed, however, he became "overwhelmed with work": "who could have foreseen such a deluge. With various help I have passed one acute stage only I fear to pass on to another." Only months later Morgan declared himself, none too happily, to be "head over ears in my flies."[8] The problem was not only the reproductive rate of fruit flies, but also their propensity to mutate in response to environmental change—the precise feature that made Drosophila so valuable to the geneticist interested in hereditary features and mutations across generations. Mendel's peas had been docile and well behaved by comparison: they were smooth or shriveled, and followed comparatively clear patterns of generational inheritance—a far cry from the seemingly endless variety of eye colors, wing shapes, and body sizes that emerged in the Drosophila "breeder reactor." In the face of this nineteenth-century data deluge, geneticists "had no choice but to adopt a fundamentally new system of naming and classifying factors."[9] In the lab, Drosophila became a new creature, one that could not exist outside that institution. But, it also reconfigured the lab itself, giving rise to new kinds of scientific places and persons, including "a new variety of experimental biologist, with distinctive repertoires of work and a distinctive culture of production"[10] In Kohler's striking language, experimental biologists became "lords of the fly," and the flies returned the favor.

Data Demand Care

Like Drosophila and Zea Mays, contemporary ecological data may be thought of as an awkward and improbable species that has nevertheless found its perfect ecological niche. Scientific data once fit on a few sheets of paper, which could last centuries if properly stored; now, we have cultivated strains of data so densely compacted they need us to take intricate care of them. As Cory Doctorow describes in a cover article for *Nature*, we have created immense industrial data centers to store and process all this scientific information.[11] In *Welcome to the Petacenter*, Doctorow stands in awe of the hundred-million-dollar computing centers that have been established to store the tens of thousands of terabytes (a terabyte being a thousand gigabytes) of data flowing from dozens of meteorological satellites, hundreds of genomic sequencers, thousands of ecological field sites, and the millions of sensors at the Large Hadron Collider. Just as the *Zea Mays* species of corn would die out in a couple seasons without our assistance, these computing centers would quickly overheat if not for the multistory cooling centers that control the massive quantity of heat they produce. If the primary, secondary, and tertiary cooling systems fail, it would only take ten minutes for the disk drives to bring their environment to a hazardous 42 °C (108 °F)—any hotter and they would begin to crack and break.

These hives of industrialized data storage are potent symbols and key infrastructures for the current era of "big" and "data-driven" science. But, the data center is also just that, the *center* of a much larger and much more complex network that extends all the way from field sites and laboratories to desktop computers at universities in every corner of the world. Push beyond the chrome exterior of the data center and you will find a squeamish student taking spit samples and delivering them to a genomics lab; scratch the silicon surface and you'll uncover a frustrated field technician recalibrating a vandalized weather monitoring station for the third time that month, or a professor pleading with a county clerk for access to the latest tax assessment records. For, as we will demonstrate, data have domesticated science not only in the sanitized environments of the industrial data center, but also at every stage, moment, and site of scientific activity. In order to support our growing appetite for scientific knowledge, we have entered into a symbiotic relationship with data—remaking our material, technological, geographical, organizational, and social worlds into the kind of environments in which data can flourish.

Behind the Data Archive

This morning I'm working on a paper and I'm looking at data and I'm making graphs, writing this paper and the graphs are swell and the statistical analysis is coming up super well. I nearly went down the hall to thank the lab crew because whenever I do this. . . .You realize how many things have to go right in order to get that graph. I mean, so we had to design the study well but then the samples had to be collected right and then they had to be handled right and they had to be extracted right and then the chemical analysis and the incubation and like, so many . . . I always enjoy that process and I always enjoy realizing how much goes into it in order for it to come out right. So, I think this is an interesting topic.
—Ecoscientist

By turning our attention to the Petacenter we came closer to the invisible infrastructures of data. Technicians, robots, and cooling systems are increasingly hidden in the clouds of computing, laboring to preserve the data of the earth sciences and, agnostically, those of many others. However, the work of sustaining massive repositories reveals only a thin slice in the long chain of coordinated action that stretches back directly to a multitude of local sites and operations through which data in their "raw" form get mined, minted, and produced. What remain at repositories are the distilled products of these field sites; behind these centers lie an even more occluded set of activities that produce those data themselves.

For the remainder of this chapter we focus on a stream chemistry data set of the Baltimore region. Ecoscientists have been collecting these data for thirteen years. Each year their data sets grow. A further trickle adds one more column: 2011, 2012, 2013. . . . Each year these data must be made commensurate with those that came before. This is how such data accrue value for scientific research.

One ecologist draws an analogy between their research approach and the practice of urinalysis during a routine medical exam: "So, you go to the doctor and the doctor samples urine and they can tell something about how the patient, the body, is functioning based on the chemistry of the urine. And if you stress the patient, those stresses are going to be reflected in the chemistry of the urine and it's the same with a watershed." Changes in the environment are reflected in the chemistry of the stream flow. As our scientists like to say: seasons follow annual cycles, but ecological change occurs over decades and centuries. One way to follow such changes and disentangle such processes is through the patient work of building data sets that match those time frames. And one way of studying *that* is to follow a similarly patient approach in observing, studying, and collaborating with the people who do such work.

How to Measure the Same River Twice

For ecologists collecting data, the age-old maxim "you can never step in the same river twice" is not a philosophical reflection, but a practical problem. It is precisely "the same river" that from week to week ecologists wish to take temperature readings and collect water samples. Data only become longitudinal if they measure the same thing week to week and year to year. And yet it is also differences in those field sites over time that are of greatest interest to scientists. When are changes the right kind of changes? And, when are they no longer measuring the same river twice? In this section, we follow the work of scientists, students, and technicians as they each perform this delicate balancing act.

Over the last thirteen years, a lot has changed, much of it beyond the control of the research team. The conditions of possibility for production of data are continuously evolving. Each change threatens to compromise the comparability of data across the years, and thus, the very enterprise of a longitudinal data set. While today our stream-flow scientists take samples from sixteen different collection sites, that number has risen and dropped over the years. Occasionally, entire streambeds have ceased to be rivers at all—drying up as water consumption changes in Baltimore. Other sites have become unviable because of neighborhood development projects or industrial activity. Sometimes sites are actively damaged or vandalized. Instruments, left behind from week to week, are stolen or left covered in graffiti. Economic conditions in Maryland contribute directly to this. Our scientists report that their instruments can be scavenged for parts or raw materials (such as copper cables). Posing additional challenges are the recent introduction of automated instruments to chemical water sampling and analysis in the Baltimore area: producing end-to-end changes in data routines and requiring months and years of painful and sometimes uncertain calibration work before the new results can be reliably matched to data produced by older techniques.

To understand the mutual construction of data and the everyday work of scientists, as well as the orientation to producing comparable longitudinal data, we cover these difficulties in three sections: (1) the weekly rituals and routines used to generate measurements that make up the database; (2) the field sites and instrumentation that both threaten and comprise the very purpose of the longitudinal study; and (3) those practices that carry data from field sites to the databases themselves. We developed these insights through ethnographic field research and from the accounts of data collectors who themselves characterize their difficulties and the lived work of data collection.

Routines and Ritual: "We Go Out on Wednesdays"

For the last sixteen years, teams of three or four ecoscientists, technicians, and graduate students have set out in a van once per week (most often on a Wednesday) to visit sixteen field sites in Baltimore county. The path is a circuit for the driver, repeated routinely. Sites are streambeds, located at driving distances of fifteen minutes apart to just over an hour. On a summer day in 2011, we join the on-site team for a day, acting as self-identified ethnographers and offering to participate in the manual labor.

The day trip has an easy feel to it, and the accompaniment of ethnographers raises no eyebrows; visitors are common. At the first signs of our expressed curiosity during an interview, the lead technician invited us to join on their trip. Our first scheduled visit to join was postponed because we competed with teachers for the two extra seats in the van. To have visiting scientists, elementary school teachers, or ecologically minded community members join is common and there is a commitment to outreach and education that piggybacks the collection ritual. When our turn comes we are warned only that we will have to be willing to get dirty and help carry a few things.

At 8 a.m. we congregate in the parking lot. Fitting ourselves from the plentiful supply of high rubber boots and spraying ourselves against mosquitoes, we are committed to a day's work. The team includes a lead technician in his mid-forties and two graduate students dressed in casually hip outdoor gear. We help load stacks of plastic bottles and a plentiful supply of fresh coffee into the trucks and head out from the University of Maryland's Baltimore County campus.

At 8:45 a.m. we arrive at our first sampling site. There is nothing in particular that visually distinguishes this first stop as a field site per se: nothing more than a road intersecting a small bridge, and a river swirling below. The driver pulls the van onto the gravelly side of the road and we disembark, bringing with us the necessary equipment: three small plastic bottles, three instruments affixed to the technician's belt, and an invaluable pen and field sheet in hand. The walk from road to river is a worn path through the underbrush and a hop into the streambed. Our boots protect our feet from the light flow of the water. We can still see the road, the passing traffic, and the neighborhood of single-unit homes around us.

The core of the collection ritual is as follows. First, we begin by inspecting a set of worn-looking gauge sticks that our scientists dug into the riverbed long ago. Each metered stick is partially immersed in the stream and a quick glance reveals the height of the water. Today, one stick is out of the water, no longer within the pathway of the river, thereby indicating a need for recalibration. On the field sheet our guides record the height of the water and note the misplaced stick (figure 8.1). Second comes sampling. Each of the three plastic bottles are dunked in the water, emptied, filled again, and then capped. One of the graduate student researchers records a series of matching numbers: one on each of the bottles, and one on the field sheet. The pen travels the short distance from bottle to field sheet, recording a matching number on each. Finally, each instrument is immersed in the stream, only to be emerged with readings for temperature, turbidity, oxygenation, and acidity. At 9:04 am we gather the bottles, check for debris, and climb back into the truck to head to the next site.

Half of the sites are located in the heart of Baltimore, nestled in residential neighborhoods and industrial zones. As the day wears on and we spiral out from the urban core of the city, the landscape and its residents change: from the dense interlocking residential neighborhoods and industrial zones of the downtown core to the lawns and open spaces of more affluent neighborhoods, and eventually to the more pastoral landscapes of state parks. Our collection sites track and reflect these variations. For instance, in the way we typically imagine these things two of the urban sites could not be considered

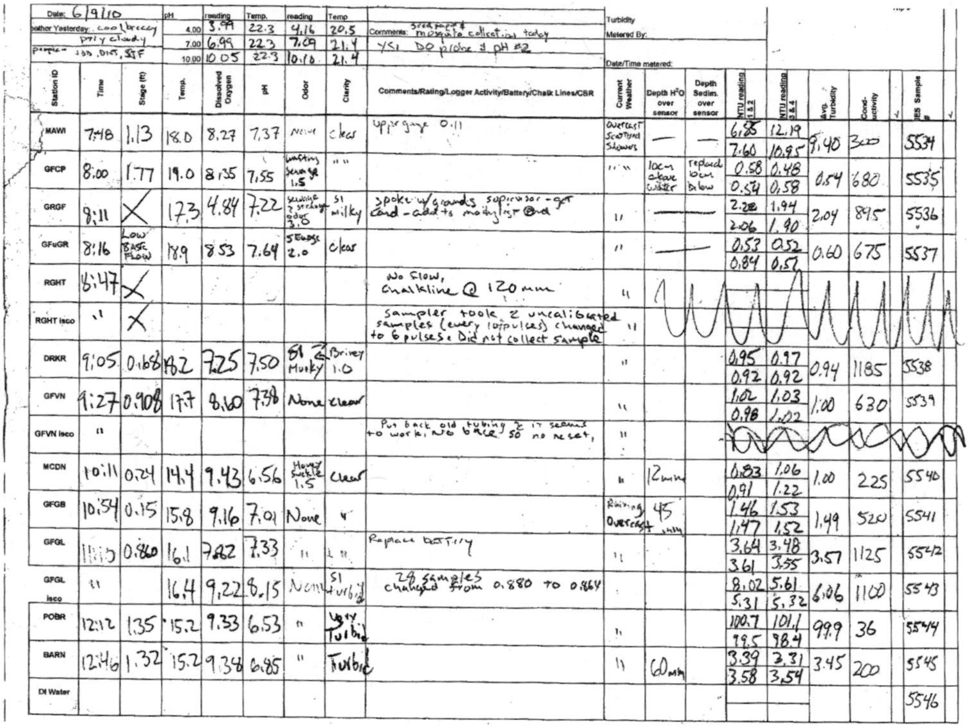

Date: 6/9/10
Weather Yesterday: cool breezy, partly cloudy

pH	reading	Temp.	reading	Temp
4.00	3.99	22.3	4.16	20.5
7.00	6.99	22.3	7.09	21.4
10.00	10.05	22.3	10.10	21.4

Comments: mosquito collection today; YSI DO probe & pH #2

Turbidity Metered By:
Date/Time metered:

Station ID	Time	Stage (ft)	Temp.	Dissolved Oxygen	pH	Odor	Clarity	Comments/Rating/Logger Activity/Battery/Chalk Lines/CSR	Current Weather	Depth H²O over sensor	Depth Sedim. over sensor	NTU reading 1 & 2	NTU reading 3 & 4	Avg. Turbidity	Cond-uctivity	ES Sample #
MAWI	7:48	1.13	18.0	8.27	7.37	None	clear	upper gage 0.11	Overcast Scattered Showers	—	—	6.85 / 7.60	12.19 / 10.95	9.40	300	5534
GFCP	8:00	1.77	19.0	8.35	7.55	wafting sewage 1.5	" "		" "	10cm clear water	replaced 10cm below	0.58 / 0.54	0.48 / 0.58	0.54	680	5535
GRGF	8:11	X	17.3	4.84	7.22	sewage 2 sewage odor 3.0	Sl milky	Spoke w/grounds supervisor - get card - add to mailing list	"	—		2.28 / 2.06	1.94 / 1.90	2.04	895	5536
GFuGR	8:16	Low BASE FLOW	18.9	8.53	7.64	sewage 2.0	clear		"	—		0.53 / 0.84	0.52 / 0.52	0.60	675	5537
RGHT	8:47	X						No flow, chalkline @ 120 mm	"							
RGHT isco	"	X						Sampler took 2 uncalibrated samples (every 10 pulses) changed to 6 pulses. Did not collect sample	"							
DRKR	9:05	0.168	18.2	7.25	7.50	Sl Briney 1.0	Murky		"			0.95 / 0.92	0.97 / 0.92	0.94	1185	5538
GFVN	9:27	0.908	17.7	8.60	7.38	None	clear		"			1.02 / 0.98	1.03 / 1.02	1.00	630	5539
GFVN isco	"							Put back old tubing & it seems to work, no battery so no reset.	"							
MCDN	10:11	0.24	14.4	9.43	6.56	Honey suckle 1.5	clear		"	12mm		0.83 / 0.91	1.06 / 1.22	1.00	225	5540
GFGB	10:54	0.15	15.8	9.16	7.01	None	"		Raining Overcast	45mm		1.46 / 1.47	1.53 / 1.52	1.49	520	5541
GFGL	11:15	0.860	16.1	7.82	7.33	"	"	Replace battery	"			3.64 / 3.61	3.48 / 3.55	3.57	1125	5542
GFGL isco	"		16.4	9.22	8.15	None	Sl turbid	24 samples changed from 0.880 to 0.864				8.02 / 5.31	5.61 / 5.32	6.06	1100	5543
POBR	12:12	1.35	15.2	9.33	6.53	"	very turbid		"			100.7 / 99.5	101.1 / 98.4	99.9	36	5544
BARN	12:46	1.32	15.2	9.34	6.85	"	Turbid		"	60mm		3.39 / 3.58	3.31 / 3.54	3.45	200	5545
DI Water																5546

Figure 8.1 A completed field sheet

streams at all: to us they appear as sewage and drainage pipes. In the exurbs, our guides tell us, the water might be filtered down from private septic systems. Visually, very little sets the sampling sites apart from their surrounding urban landscape. Some of the sites are marked by discrete metal boxes containing automated sampling equipment, but to our untrained eyes these pass as electrical infrastructure.

Traveling in a dusty van piled with equipment, our ecoscientist team spends most of the day together—stopping occasionally to collect samples of water, take temperature measurements, and share a meal. The ritual has been repeated thousands of times, but no single practical or material element endures the years: students graduate to faculty, instruments become outdated or imprecise, even buckets wear out. We cannot even say that it is a routinized practice that persists, for that has been modified to fit novel instrumentation, changes in the sites of collection, or the execution of new subprojects in data collection.

We can only claim that the ritual is the same in the sense that we can claim that each time our scientists collect stream data they are stepping in the same river twice. What persists across each iteration of the ritual is the *comparability* of the data collected; it is the purpose and orientation of their activity. This comparability is the achievement of carefully coordinated effort that stretches out every week to the sixteen field sites and back to the labs at the University of Maryland.

In the language of anthropology, what differentiates routine from ritual is the meaning for participants. Routines are dry and mechanical. While routines are always adapting to local circumstance, changes make little difference to those involved. Rituals are lived. They may be enjoyable or tedious, but rituals are experienced as a feature of member-ship. Rituals tie activities to a past, and through enactment, reproduce that past into the present and future. The continuance of the data set is what sets the activities of these ecoscientists apart from routine.

An Orientation to Comparability

Arrival at each geographic site is initiated with a quick visual inspection for discrepan-cies: Is anything notably out of place? Are there higher or lower flows of stream water, residues of flooding, evidence of a disturbed sampling machine, graffiti on a bridge, or a strange smell? Such observations become the first raw data collected at each site, qualitatively recorded on the field sheet.

Four artifacts leave each field site: a field sheet and three bottled samples of stream water. The field sheet is a single-sided piece of paper; it begins each week as the same empty chart and ends each collection day with the qualitative and quantitative inscrip-tions recording observations for each of the sixteen sites. It is the documentary trace of the day's work. The samples of water only become data later; one of these bottles is processed in labs at the University of Maryland at the end of the day while the other two bottles travel to Milbrook, New York, for analysis weeks later. The top of the field sheet is analogous to the start of the collection day: it begins by documenting the date, the data collectors, and qualitative notes about the weather. The next step is calibration of the instruments, checking their accuracy against standardized acid and temperature metrics available in the labs at Maryland.

At the riverbed, the field site itself becomes data in front of our eyes through a practice of observation and a set of practical interventions. Smell is evaluated at each site and recorded on the field sheet: terse but florid descriptions accompany a number between 0 and 4: "pickles and propane, 3," "no smell, 0," "benzene (which is a whisky-like smell), 4." Samples are collected in small plastic bottles that are first filled and then emptied into the stream to be sampled—thus, clearing any residue from last week's ritual. The practical activities of data and sample collection are technical, but not esoteric. By the fourth site, we visitors were invited to collect temperature samples or hold the field sheet, filling in the called-out measurements. To take a temperature reading, we had to wade into the middle of the water and hold out the thermometer upstream of our bodies.

The routine is simple and quickly learned, but experience teaches one how to manage outliers. If the smell of the field site is toxic ("methane," "chemical," "disgusting"), the site may be evacuated immediately, leaving only that trace recorded on the sheet. A single failed reading in the field sheet (what eventually becomes a blank entry in the database) does not threaten the comparability of the data; it is only over time that this failed reading becomes a concern.

During collection, participants are familiar with aligning a past set of data-capture activities with the circumstances presented to them at each field site; it is a form of standardization oriented to sustain alignment with past measurements. Observation and documentation at each site are focused on detecting changes relevant to the commensurability of past versions of the ritual. Such changes are meaningful in that they simultaneously threaten the data set and promise new developments in knowledge.

Each step in the activities of collection is routine and standardized. In this sense, the steps are mundane. Nevertheless, each activity is conducted with an orientation to comparability and managing situational differences. Differences are judged meaningful through activities of observation and made accountable through discussions between scientists conducted in situ at each site.[12] The database, the full archive of recordings stretching back sixteen years, is not physically present in the ritual. It is even likely that some of the technicians have never so much as glanced at the database. Yet, in the routinized activities of data collection, and in the perceptual orientation generated by the empty boxes of the field sheet, a concern with comparability (with that very database) is fostered.

Shifting Field Sites: Environment, Humans and Infrastructure over Time

For scientists, change in the field sites is the name of the game, but too great a change and these sites cease to be relevant at all. Change is both the source of new knowledge and an incipient threat to the comparability of longitudinal data. Determining whether a site is to be considered threatened is rarely a matter of on-site decision making—it does not occur as an instant in the data collection ritual. Rather, it occurs over a period of many site visits, as the scientists begin to notice a pattern week after week in their collection activities, and as the descriptive metadata on the field sheet pile up. Have environmental conditions systematically challenged data collection? Has new industry created a high point of pollution that cannot be considered representative of the environs? Chemicals are what interest our scientists most, but if a factory is built too close to a sampling site then the data are not generalizable. The environment itself is not considered static—transformations are expected. How do you know whether a change is revealing or compromising? In the section that follows, we step back from our ethnographic immersion to look upon past incidents in the oral history of the data set that have threatened its viability.

Environmental Changes: "There Is No Water to Measure"

In 2002, Baltimore was subject to a record drought. This drought caused visible transformations to the urban and forest ecologies within the county. Our researchers found many of their streambeds completely dried up. With no water to sample and no temperature to measure for months on end, little information was recorded in the field sheets.

The metonymy of "urinalysis" breaks down when the body of the environment cannot be read from its fluids. Stream flow can be reported as a zero, a finding in itself, but with no accompanying samples there is no chemistry to analyze in labs. As such, nothing can be reported at all in those fields of the database. However, the situation reversed radically in 2003 and 2004 due to reports of record moisture and renewed flow in Baltimore streams. The term for this reversal is a "climate pulse." These "pulses" are precisely the kinds of changes our scientists hope to examine in a longitudinal study. A short-term study, months to years, could be ruined by the inability to collect samples, but in a long-term project such pulses became data in analysis that stretched across decades.

Human Changes: A New Sewage System

In 1999, the City of Baltimore Department of Public Works (DPW) entered into a consent decree with the Environmental Protection Agency (EPA) to address sanitary and combined sewer overflows across the entire city. In short, Baltimore has experienced a population collapse over the last few decades. From the perspective of sewage, this decline presents unique "scaling down" challenges for the city's infrastructure: a system designed for a citizenry of over a million people quickly came to support less than six hundred thousand.

Our stream chemistry researchers have mixed feelings regarding this transformation. On the one hand, urban ecologies are of great interest to them: the effects of "coupled systems" (natural and human changes) is precisely what they seek to study. On the other hand, such large-scale interventions present a virtually uncountable number of variables to manage in their studies: population, demographics for that population, policy and legal interventions leading to engineering overhauls, and of course innumerable changes to the sewage system itself. For some, entire trajectories of investigation had to be scrapped. For others this presented a natural experimental condition: "This, for us, was a great experimental opportunity because we had seven years of background data off these twelve streams and a few of them were very strongly affected by the sewage improvements and a few of them were not."

Considered as an "experiment," the new sewage system provided a unique occasion for a novel study that no other researcher has had the ability to enact. Long-term data stretching before and after a change will open a window of understanding on urban renewal. Many cities in America and around the world are going through a similar process. But, how are these new data to be reconciled as a single longitudinal arc? Scores of variables that were well understood are thrown into a complex flux—making environmental claims difficult for those scientists to assert.

Instruments: Breakdown and Automation

In a longitudinal study instruments come to be part of the field sites themselves. At each of the sixteen sites, meter sticks are strategically placed in the streams. These sticks are dug into the ground on metal poles or affixed to the walls of overpasses. These allow for quick and standardized gauges of the height of the water flow, on each occasion measured from the same location. However, water flows are not static—by their very nature they continuously dig away at their own streambeds. As one scientist noted: "Sometimes our poles stop being in the water at all. That's when we realize that our readings might have been off for some time. That's a pain: we'll have to adjust recent data, and figure out where to put the gauge meters next."

Local residents are sources of consternation, as they interfere with instruments, sometimes in ways that make it difficult to know this even happened. Each site has a rain meter—a small open pipe that fills with rainwater—providing a measurement of rainfall; these "pipes seem to be an irresistible temptation for kids to pee in." There is always the urge to find designs that avoid such human interference or to increase the instruments' precision, thereby reducing labor through automation. But, each such improvement in infrastructure inevitably presents a challenge to the record and, thus, to the sustainability of the long-term endeavor. While a gauge meter is crude and occasionally needs calibration (e.g., reaffixing the stick), it is also very reliable, easily available, and requires very little user expertise. Our scientists are conservative toward their instruments, protocols, and objects of study. They do not add a new chemical test to the repertoire as it becomes available without an assurance that the test will remain available, affordable, and able to keep measuring "the same thing" across the years.

These ecologists fight a three-front war with their closest allies. In order to sustain a comparable archive, data demand the taming of unruly field sites, humans, and infrastructure. A dance of stability and change emerges in an ongoing effort to isolate environmental transformations that can stand in for something broader than a streambed.

Cascading Rituals: From Field to Lab

We dived from the warm petacenter directly to the flowing field site, revealing the background work of creating comparable long-term data. Thus far, we have focused on how participants are oriented to aligning a past data archive with a present practice of data collection. Differences found at any given site—whether environmental, human, or infrastructural—are subjected to a test of relevance: is the ritual of collection threatened, and in turn, will changes present difficulties to comparability with past data? This is an accurate description of the orientation of participants *as they go about the task of collecting data and samples on site*; nonetheless, it leaves all the space between the archive and the field site unaccounted for.

Only a small portion of what comes to be considered raw data is actually generated in the field. Specialized tools, such as thermometers and meter sticks generate quantitative results, such as temperature and gauge, while, moderate training of the senses produces qualitative perceptions, such as smells. With only these mediators, our ecoscientists have for over a decade transformed the natural world into data on site. Bruno Latour has described these as the moments whereby matter becomes form, where some of the materiality of streams is sloughed off in favor of greater mobility of data.[13] Transformed into numbers and writing on field sheets, these mobile facts can be easily ported back for data entry. But, the greater part of *becoming raw data* only comes to pass further down a chain of mediations: in the laboratories of Maryland or Milbrook.

Producing those data means isolating and transporting little bits of streams back to labs in ways that preserve meaningful relationships to those streams. These bits are called "samples": a straightforward term that belies the work that meaningfully sustains them as representing a stream at a particular point in time. How is the relationship between a field site and its sample preserved? This is a mundane question, as are the methods used to transport and coordinate those samples. On the right side of the field sheet is a number that facilitates the movement of the sample from field site to lab: the "sample #." The same number is placed on the sample bottles: once on the bottle itself and once on the lid. It is this number that holds together the relationship among a date, a field site, and the bits of water that are trucked away.

Many of these bits of river continue on for years as samples rather than data, preserved in massive cold rooms in the basements of Cary Institute of Ecosystem Studies in Milbrook. Every month, the samples for several weeks of ritualized work are loaded onto a chilled truck that travels from Baltimore to Milbrook. The samples are then carefully ported to these cold rooms. This physical archive preserves samples of Baltimore streams, stretching back almost two decades, along with bits of other bodies of water that go back even further. In the face of a new laboratory technique or scientific question, these samples could be used to regenerate an entire new data stream stretching back as long as the numerical archive. It is this simple alignment, a number on a field sheet matched to two on a bottle, that make accountable the representativeness of each sample for decades, and possibly centuries. If this simple numbering ritual fails, so too does the chain that connects the field to the lab: "That's one long-term study we haven't done. The life of the label glue over time. I dread to think that one day we'll walk in to the cold storage room and hundreds of labels will be lying scattered beside the bottles. But I think the extra label tape we put on the lids will hold up."

These sample numbers are a kind of data that never make it into the final archive. They are used to coordinate the movement of samples across physical and temporal distances, after which they are discarded. We could call these numbers procedural metadata. Metadata, as the meme goes, are data that describe data. Usually, we think of these as contextual information: the date, time, and location of a measurement, who took the reading, when was the instrument last calibrated. These kinds of metadata can be used to understand and evaluate data at later points in time, or used by those unfamiliar with the collection rituals. But, procedural metadata serve only in the interim periods, as samples are transported. A routinized check comparing samples to field sheets occurs at each end of each trip: from field site to Baltimore labs, or from Maryland to Milbrook.

There are others bits of the river that are shed along the way, never making it to the main database. For instance, conductivity is another measure taken right at the field site. This measure is recorded on the field sheets in the second to last column, but it travels no further along the chain. Conductivity accompanies our scientists back to the lab, but there it is forgotten, or rather, buried in a mound of archived field sheets: "It is possible to go back to these data sheets if necessary, but they would have to dig." When we asked why these data points made it no further, the glib response was simply that no entry existed for conductivity in the database. Meanwhile, well-worn columns of the database were filled with qualitative observations about smells and random field events.

This labeling ritual, the notations on samples, the checks at each point of transport, are the cascades of rituals that tie together field sites to samples to databases. We have only scratched the surface of these events. Our scientists described how samples are placed in the car just to prevent overturning. Bottles, whether filled or empty, are transported with sealed lids to prevent cross-contamination. Shifting the contents of water from one bottle to another or to an instrument involves isolation from other samples to ensure none are confused. We observed cascades of rituals, from the moment of a sample's collection in the river to its placement in the lab refrigerator.

At each of these tiny transitions data again threaten to become unruly masses. A misinscribed sample number, a confusion of two bottles, or the spilling of a sample during filtration can all threaten the chain that links a date and a field site to a sample and its eventual transformation into data. A myriad of ritualized activities seek to solidify this chain, but small mistakes and accidents still occur. At best, a mistake or accident is caught and a data point is lost. While a single data point is a loss, in the grand scheme of a longitudinal database, it is a fairly small one. Scientists who use these data expect such things: anomalies and outliers that must be thrown out, missing data points that can be interpolated or extrapolated. At worst, a mistake becomes systematic (as with a misplaced gauge stick), whereby entire sections of a data stream must be reconstructed or altogether thrown out.

The metaphor of a chain is revealing: it helps us understand the heterogeneous work of custodianship stretching from field site to lab and from lab to databases. Only at the end of these mediations can we meaningfully speak of "raw data." Nevertheless, the chain is also obfuscating—implying a beginning and end for data. The archive for this chemistry stream flow is not singular; rather, it is quite literally distributed across databases, field sheets, and a physical archive of samples. When asked "Where is the stream data archive?," a researcher will first insistently direct us to a public online page with an embedded web service. But thereafter, with only a little further prodding from the interviewer, the database becomes multimedia: it is digital; it is paper and pen; it is water. The digital database is its public face, accessible around the world. It signals the presence of an archive to scientists. In the field sheets, salinity data remains silently annotated, awaiting their analyst, and in the water archive, a promise of future discoveries. All three of these, and multiple other components, make up the archive of raw data.

Conclusion

We tell ourselves that we live in an era of aggregation and automation. From this perspective, raw data patiently await assembly: potable water, environmental damage, or climate change? Click. Shuttled from data storage to a computing center, the analytical engines of the twenty-first century assemble statistics, graphs, and ever more clever visualizations in response to these and many other questions we have not yet thought to ask.

But there are stories *behind* these stories. What we have offered here is another narrative, one of temperamental and delicate creatures, whose existence and fraternity with one another depend on a complex assemblage of people, instruments, and practices dedicated to their production, management, and care. Like corn and flies before them, data demand and build the human, organizational, and infrastructural worlds around them—enforcing a burden of care and work that disappears beneath (but ultimately, constitutes) the futuristic possibilities of the petacenter. Where then does raw data begin and end? If such a clear and objective dividing line exists, we have not yet found it.

We have cast corn, flies, and petacenters as the surprising conquerors of their environments, demanding suitable treatment from their human coinhabitants. In comparison, the practical collection of water samples may seem local and mundane, but it is at this level of granularity that data exert a continuous force, on a weekly basis, requiring a new round of collection and care. It is these local collection and data practices, combined with similar practices around the world that make seeing large-scale and long-term phenomena possible. They are the very stuff of global knowledge. With only a little push from the interviewer our informants agree. A scientist worth his or her mettle never stops their investigation at publically available data. This is not where the archives of raw data reside. Rather, the archive's borders stretch to a receding horizon that include the pen and paper field sheets backfiled for years, a cold room of samples, and the uncaptured experience of scientists and technicians entrusted with the production of the archive. The field sheets are not paper relics, but rather, a source of data awaiting their user. The samples are a promise of a renovated archive, in line with the newest analytical techniques. The field techs standing knee-deep in the pickle-smelling waters of exurban Baltimore are not the opposite of global knowledge—they are *participants* in its assembly.

Acknowledgments

We would like to thank our collaborators in field research and conceptualization of this chapter: Stuart Geiger and Mathew Burton. We also thank Jessica Beth Polk for her careful and observant eyes.

Notes

1. This is a term we develop by analogy to Foucault's "commodity fiction of power," in which power is modeled as an undifferentiated force, reserve, or entity separable from specific contexts of action. The commodity fiction of data performs the same trick on data, assuming or projecting a world where data floats free of its origins, shedding form, substance, and history, and is thereby rendered free to travel the world as an undifferentiated and universal currency; but as the chapters in this volume make clear, data is stickier than that (Michel Foucault, "Two Lectures," in *Power/Knowledge: Selected Interviews and Other Writings, 1972–1977*, ed. Colin Gordon [New York: Pantheon Books, 1980]).

2. C. Thompson, *Making Parents: The Ontological Choreography of Reproductive Technologies* (Cambridge, MA: MIT Press, 2005).

3. M. Pollan, *The Omnivore's Dilemma* (New York: Penguin Books, 2006).

4. Ibid., 90.

5. Ibid., 64.

6. W. Cronon, *Nature's Metropolis: Chicago and the Great West* (New York, London: W. W. Norton & Company, 1991).

7. R. E. Kohler, *Lords of the Fly: Drosophila Genetics and the Experimental Life* (Chicago: University of Chicago Press, 1994).

8. Ibid., 47–48.

9. Ibid., 61.

10. Ibid., 49.

11. C. Doctorow, "Big Data: Welcome to the Petacentre," *Nature* 455, no. 7209 (September 4, 2008): 16–21.

12. C. Goodwin, "Professional Vision," *American Anthropologist* 96, no. 3 (1994): 606–633.

13. B. Latour, *Pandora's Hope: Essays on the Reality of Science Studies* (Cambridge, MA: Harvard University Press, 1999), see chap. 2.

Data Flakes: An Afterword to "Raw Data" Is an Oxymoron

Geoffrey C. Bowker

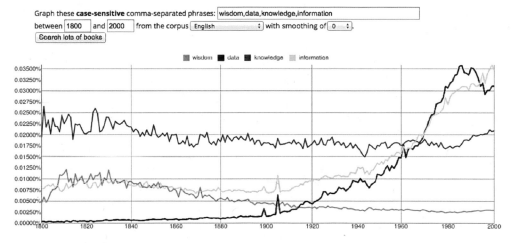

Figure 9.1   Google Ngram (http://books.google.com/ngrams) showing data and information on historic rise and knowledge and wisdom in historic decline. *Note:* Information and data peak in the late twentieth century (data is the darker line); wisdom and knowledge are in gradual decline.

Let me be hyperbolic and assert that we are entering into the dataverse. "Entering" is a key word here—it is through the labors of millions of sensors, click-workers and of course our collective selves that we are being entered.

It has been a longer-term process than most would have thought, before they read this marvelous volume. It has also been ineluctable. Harry Harrison imagined "the stainless steel rat" who could continue to swarm in worlds of concrete, glass, and cameras as privately and obscurely as the veriest ant.[1] However, it is now difficult to walk the streets of a major city without having one's progress captured by some hidden gaitkeeping device. It is getting difficult for trees to fall in forests without nonhuman observers. Data about me are stored in thousands of virtual locations (and not just because it's me . . .). As that data is reworked, processed through an online algorithm or spat out to somewhere and somewhen to the computer screen of a vigilant operator, my possibilities for action are being shaped.

So what's the story with data? The concept "raw data" can be aligned with Lévi-Strauss's use of the term "raw." When he wrote *The Raw and the Cooked*, he was describing a vast mythological set originating analytically in South America among the Bororo and spreading throughout the world.[2] His argument was that a series of binaries characterized this set, many of which were a variant of what we would call the nature/society divide. The natural was the raw (honey) and the social was the cooked (ashes). The interplay between the two structured many of the myths he was describing, using topological transformations such as, most notably, inversion. This constructed divide has been central to many societies up to the present day (though it's never fully completed—as witness ecologists arguing against the exclusion of human activity from "nature").

The divide has been politically and philosophically powerful. With a "natural" thoroughly separate from us, we can learn lessons from the book of nature and apply them deliberately to our own species. An absurdist statement of this move is given by the Natural Law Party—a mix of transcendental meditation and benign autocratic practice. A scientific guise for our times has been sociobiology—if all animals have a territorial imperative, then so must we. The sleight of hand (discussed by Bruno Latour in *We Have Never Been Modern*) which permits this appeal to the natural to be true is that our own understandings of nature project our views of ourselves.[3] Beehives in nineteenth-century Britain had kings, because it was believed that only a male could undertake the complex tasks of government (the titular monarch Queen Victoria notwithstanding). Our knowledge professionals see selfish genes because that's the way that we look at ourselves as social beings—if the same amount of energy had been applied to the universality of parasitism/symbiosis as has been applied to rampant individualistic analysis, we would see the natural and social worlds very differently. However, scientists tend to get inspired by and garner funding for concepts that sit "naturally" with our views of ourselves. The social, then, is other than the natural and should/must be modeled on it; and yet the natural is always already social.

Database development has followed this vein. The early databases were hierarchical—you needed to go down a detailed line of authority each time you wanted to retrieve a datum. Then we had relational databases, where there was still central control but much more flexible access (the database system, like society at the time, was seen as a fixed structure). Today we have moved into a world of object-oriented and object-relational databases, in which each data object lives in a Tardean paradise—any structure can be evanescent providing we know the inputs or outputs of any object within it. So databases have recapitulated social and organizational developments. And many organizations changed in the 1990s and 2000s in an effort to become more "object-oriented"; forgetting that the first object-oriented language (Simula, a precursor to Smalltalk) attempted to model work practice. Along the way, we have conceived ourselves and the natural entities in terms of data and information. We have flattened both the social and the natural into a single world so that there are no human actors and natural entities but only agents (speaking computationally) or actants (speaking semiotically) that share precisely the same features. It makes no sense in the dataverse to speak of the raw and the natural or the cooked and the social: to get into it you already need to be defined as a particular kind of monad.

The best description of the dataverse from within precipitates nicely from the wonderful observation from Benjamin that Krajewski cites in chapter 6 (this volume): "And even today, as the current scientific method teaches us, the book is an archaic intermediate between two different card index systems. For everything substantial is found in the slip box of the researcher who wrote it and the scholar who studies in it, assimilated into its own card index." We typically conceive of knowledge as passing from knowledge worker to knowledge worker via the intermediary of the datum. However, as Marx displayed so brilliantly with his M-C-M (money-commodity-money) cycle, we can achieve analytic purchase by looking at C-M-C (which in our era, felicitously, may refer to computer-mediated communication).

We can start perhaps by refining the terms of the cycle. Much of our "knowledge" today surpasseth human understanding. Stephen Hawking, in his inaugural lecture for the Lucasian Chair of Physics at Cambridge—once held by Newton, who had all those giants standing on his shoulders—pointed to the day when physicists would not understand the products of their own work.[4] With the world of string theory upon us, it is clear that we cannot "think" in the necessary 10+1 dimensions and the complex geometries they entail. Fields such as climate science or any others that deploy agent-based modeling systems are much the same. The intelligent citizen cannot read the programs that run our data sets; they can be "groundtruthed" to some extent—though increasingly scientific models are compared primarily against other models. So let's take the unnecessary human out of the equation and talk about the program-data-program or data-program-data cycles.

There is of course an ongoing relationship with the real world and the human observer (nature and society), however it is a difficult one to express. Both the natural world and its human observers are being ever more instrumented with intelligent machines. Staggering arrays of sensors and cameras furbish "us" with terabytes of data a day about the natural world and about our social activities. The "quantified self" movement is an oddly worshipful effort to celebrate this quantification (computers do not deal with "soft" data). The qualified self seems to be slipping out of the picture—the interpretative work is done inside the computer and read out and acted on by humans.

A dark vision is that our interaction with the world and each other is being rendered epiphenomenal to these data-program-data cycles. If it's not in principle measurable, or is not being measured, it doesn't exist. Thus in the natural world, we have largely as a species elected to take the quantifiable genome (https://www.23andme.com) as the measure of all life: when we save species (in seedbanks for example), we are saving irreducible genetic information—not communities (despite the fact that every individual comes with its own internal flora and fauna central to its survival; and that each individual can be understood equally as the product of a network of relationships). Collectivities that are not being measured and modeled are preserved, if at all, only accidentally. As people we are, in Olga Kuchinskaya's memorable phrase, becoming our own data. Mental disorders are less complexes than strings of measurable effects. By making them data, response regimes can be tested and implemented. However, this does not mean that completely different understandings of these disorders are not right—just that the complex, tight coupling between machines in the clinical and insurance industries and in administration entails that in order to survive in the world, we need to be able to become data within a highly ramified system. If you are not data, you don't exist: and just like the unfortunate Doc Daneeka in *Catch 22*, it doesn't really matter how often you declare yourself alive. In the old days of science studies, we used to worry about the supremacy of the hard sciences, which we rightly tied to the absolutism of the Christian theology they supplanted. Now we risk being in the grip of hard data.

This playful, insightful book offers a rather gentler path forward. If data are so central to our lives and our planet, then we need to understand just what they are and what they are doing. We are managing the planet and each other using data—and just getting more data on the problem is not necessarily going to help. What we need is a strongly humanistic approach to analyzing the forms that data take; a hermeneutic approach which enables us to envision new possible futures even as we risk being swamped in the data deluge.

I recently stumbled across this marvelous headline: "Data sharing revolution letting scientists use any database in the world instantly."[5] The author of this breakthrough tells us that ontology is the solution: "Ontology is philosophy. It is underlain by a philosophical system that has been unbroken since the time of the ancient Greek philosopher Aristotle (BC 384–322). It took five years for us to be able to understand information technology based on a philosophy that has been nurtured over such a long historical period." Such a terrible loss of materiality and praxis. It is certainly true that computer scientists have tried to commandeer ontology.

The authors in this volume suggest that phenomenology is a good practice to guide our understanding of and ultimately to transform this overweening vision. Computers may have the data, but not everything in the world is given.

Notes
Thank you to Judith Gregory for a generous and creative reading of this text.

1. Harry Harrison, *The Stainless Steel Rat* (New York: Walker, 1970).

2. Claude Lévi-Strauss, *The Raw and the Cooked* (New York: Harper & Row, 1969).

3. Bruno Latour, *We Have Never Been Modern* (Cambridge, MA: Harvard University Press, 1993).

4. Stephen W. Hawking, *Is the End in Sight for Theoretical Physics?: An Inaugural Lecture* (Cambridge, UK: Cambridge University Press, 1980).

5. See http://www.labonline.com.au/articles/50252-Data-sharing-revolution-letting-scientists-use-any-database-in-the-world-instantly.

Contributors

Chapters

Geoffrey C. Bowker is a professor at the School of Information and Computer Science, University of California at Irvine, where he directs a laboratory for Values in the Design of Information Systems and Technology. Recent positions include professor and senior scholar in cyberscholarship at the University of Pittsburgh iSchool; and executive director, Center for Science, Technology, and Society at Santa Clara University. Together with Leigh Star he wrote *Sorting Things Out: Classification and Its Consequences* (MIT Press, 1999); his most recent book is *Memory Practices in the Sciences* (MIT Press, 2005).

Kevin R. Brine is an independent scholar and artist with a background in investment management and finance. Formerly a partner and board member of Sanford C. Bernstein & Co. Inc., Brine has extensive experience on both for-profit and nonprofit governing boards. He is currently coauthoring with Mary Poovey a book-length study of the social and intellectual conditions that led to the 2008 global economic crisis, provisionally called "A Model for the Future: A History of American Finance Capitalism."

Ellen Gruber Garvey is the author of *Writing with Scissors: American Scrapbooks from the Civil War to the Harlem Renaissance* (Oxford University Press, in press) and *The Adman in the Parlor: Magazines and the Gendering of Consumer Culture* (Oxford University Press, 1996). Her recent articles include work on women editing periodicals, and on the advertising of books. She is professor of English at New Jersey City University.

Lisa Gitelman teaches English and media studies at New York University. She is the author of *Always Already New: Media, History, and the Data of Culture* (MIT Press, 2006) and coeditor with Geoffrey Pingree of *New Media, 1740–1915* (MIT Press, 2002).

Steven J. Jackson is an assistant professor of Information Science at Cornell University. He works broadly in the areas of science and technology policy, scientific collaboration, democratic governance, and global development, with special emphasis on practices of collaboration and governance in collective knowledge activities (where governance = policy + organizations + informal norms and expectations + the shaping influence of material, computational, and natural environments). He also studies how infrastructure—social and material forms foundational to other kinds of human action—gets built, stabilized, and sometimes undone.

Virginia Jackson is the UCI Endowed Chair in Rhetoric and Communication at the University of California, Irvine. Her first book, *Dickinson's Misery: A Theory of Lyric Reading* (Princeton University Press, 2005) won the MLA First Book Prize and the Christian Gauss award that year. Her second book, *Before Modernism*, is forthcoming (Princeton University Press). With Yopie Prins, she has coedited *The Lyric Theory Reader* (Johns Hopkins University Press, forthcoming).

Markus Krajewski is an associate professor of media history of science at the Faculty of Media at Bauhaus University Weimar. In 2008–2009 he was a fellow at the Humanities Center at Harvard University where he also taught as a visiting professor in the Department of the History of Science. He is the author of *Paper Machines: About Cards & Catalogs, 1548–1929* (MIT Press, 2011); *Der Diener: Mediengeschichte einer Figur zwischen König und Klient* (S. Fischer, 2010), translated as *The Servant: Media History of a Figure between King and Client* (Yale University Press, forthcoming); and of *Restlosigkeit: Weltprojekte um 1900* (S. Fischer, 2006). He is also developer and maintainer of the bibliography software *synapsen—a hypertextual card index* (http://www.verzetteln.de/synapsen).

Mary Poovey is the Samuel Rudin University Professor in the Humanities and professor of English at New York University. She has published widely on subjects ranging from feminist theory to the history of accounting. Her most recent book is *Genres of the Credit Economy: Mediating Value in Eighteenth- and Nineteenth-Century Britain* (University of Chicago Press, 2008). Her current work, coauthored with Kevin R. Brine, focuses on the history of financial modeling.

Rita Raley is an associate professor of English at the University of California, Santa Barbara, where she teaches courses in the digital and global humanities. She is the author of *Tactical Media* (University of Minnesota Press, 2009) and has recently published articles on topics such as locative narrative, code art, and text-based media installations.

David Ribes is an assistant professor in the Communication, Culture & Technology program (CCT) at Georgetown University. He studies the emerging phenomena of cyberinfrastructure (i.e., networked information technologies in the support of science) and how these are transforming the practice and organization of contemporary knowledge production. A common theme of his research is investigating the sustainability of long-term research organizations. His primary methods are ethnographic and archival. For more information, see http://davidribes.com.

Daniel Rosenberg is a professor of history at the Robert D. Clark Honors College at the University of Oregon and editor at large at *Cabinet: A Quarterly of Art and Culture.* With Anthony Grafton, he is author of *Cartographies of Time: A History of the Timeline* (Princeton Architectural Press, 2010).

Matthew Stanley teaches and researches the history and philosophy of science. He holds degrees in astronomy, religion, physics, and the history of science and is interested in the connections between science and the wider culture. He is the author of *Practical Mystic: Religion, Science, and A. S. Eddington* (University of Chicago Press, 2007), which examines how scientists reconcile their religious beliefs and professional lives. Currently, he is writing a book that explores how science changed from its historical theistic foundations to its modern naturalistic ones. Professor Stanley is also part of a nationwide National Science Foundation-funded effort to use the humanities to improve science education in the college classroom. He has held fellowships at the Institute for Advanced Study, the British Academy, and the Max Planck Institute.

Travis D. Williams is an assistant professor of English at the University of Rhode Island. His research concerns the intersections of literature, rhetoric, and mathematics in early modern Britain. He is a coeditor of *Shakespeare Up Close: Reading Early Modern Texts* (Arden Shakespeare).

Captions and Color Plates

Thomas Augst is associate professor of English at New York University. He is the author of *The Clerk's Tale: Young Men and Moral Life in Nineteenth-Century America* (University of

Chicago Press, 2003), and coeditor of *Institutions of Reading: The Social Life of Libraries* (University of Massachusetts Press, 2007).

Jimena Canales is an associate professor at the Department of the History of Science at Harvard University and author of *A Tenth of a Second: A History* (University of Chicago Press, 2010). She specializes in the history and philosophy of the physical sciences, focusing on epistemology, science and representation, and theories of modernity and postmodernity. Her published work covers the history of architecture, film, relativity theory, and nineteenth- and twentieth-century science and philosophy. Among her recent articles are "Desired Machines: Cinema and the World in Its Own Image," *Science in Context* (2011); "A Science of Signals: Einstein, Inertia, and the Postal System," *Thresholds* (March 2011); and "'A Number of Scenes in a Badly Cut Film': Observation in the Age of Strobe," in *Histories of Scientific Observation*, ed. Lorraine Daston and Elizabeth Lunbeck (University of Chicago Press, 2011).

Paul E. Ceruzzi is curator of Aerospace Electronics and Computing at the Smithsonian's National Air and Space Museum in Washington, DC. He received a BA from Yale University and a PhD from the University of Kansas. At the Smithsonian, he has curated a number of exhibits concerning the interplay of computing and aerospace technology. He is the author of several books on the history of computing, including *Reckoners: The Prehistory of The Digital Computer* (Greenwood Press, 1983), *Beyond the Limits: Flight Enters the Computer Age* (MIT Press, 1989), *A History of Modern Computing*, 2nd ed. (MIT Press, 2003), and *Internet Alley: High Technology in Tysons Corner* (MIT Press, 2008).

Ann Fabian teaches American history at Rutgers University. Her most recent book is *The Skull Collectors: Race, Science, and America's Unburied Dead* (University of Chicago Press, 2010).

Lisa Lynch works broadly at the intersection of culture, technology, and political change. Her research areas include emerging media and web culture, the changing practices of journalism, post-Cold War nuclear culture, and human rights. From 2004 to 2006, she was the director, along with Elena Razlogova of the Guantanamobile Project, a multimedia documentary about the U.S. detention of prisoners at Guantanamo Bay. Her work has appeared in publications ranging from *Literature and Medicine* and *New Literary History* to *Open Democracy* and the *Arab Studies Journal*. Her most recent research project funded by the Social Sciences and Humanities Research

Council of Canada explores what Canadian media organizations understand about Internet infrastructure.

Lev Manovich (http://www.manovich.net) is a professor in the Visual Arts Department at the University of California, San Diego (UCSD), and director of the Software Studies Initiative at California Institute for Telecommunication and Information (Calit2). *Jeremy Douglass* is assistant professor of English at the University of California, Santa Barbara (UCSB). *William Huber* is a PhD candidate in the Visual Arts Department at UCSD.

Vikas Mouli is a private equity investor with TPG Growth, based in San Francisco. Previously, he worked at Barclays Capital in the Mergers & Acquisitions Group, focused on the energy sector. Mouli graduated with honors from Harvard, with a BA in economics and history of science.

Index